Creative Intelligence in Mathematics and Science

Other World Scientific Titles by the Author

Deep Thinking: What Mathematics Can Teach Us About the Mind
ISBN: 978-981-4618-03-8 (pbk)

Creative Intelligence in Mathematics and Science

William Byers
Concordia University, Canada

World Scientific
NEW JERSEY • LONDON • SINGAPORE • BEIJING • SHANGHAI • HONG KONG • TAIPEI • CHENNAI • TOKYO

Published by

World Scientific Publishing Co. Pte. Ltd.
5 Toh Tuck Link, Singapore 596224
USA office: 27 Warren Street, Suite 401-402, Hackensack, NJ 07601
UK office: 57 Shelton Street, Covent Garden, London WC2H 9HE

British Library Cataloguing-in-Publication Data
A catalogue record for this book is available from the British Library.

Creative Intelligence in Mathematics and Science

ISBN 978-981-98-1913-3 (hardcover)
ISBN 978-981-98-1914-0 (ebook for institutions)
ISBN 978-981-98-1915-7 (ebook for individuals)

For any available supplementary material, please visit
https://www.worldscientific.com/worldscibooks/10.1142/14479#t=suppl

Preface

The subject of this book is creative intelligence in mathematics and science. What is creative intelligence? I don't have an easy answer to provide. In one sense, we all have some feeling for the meaning of intelligence and the meaning of creativity, but both of those notions are notoriously hard to pin down. Sometimes when you reflect on the meaning of these terms, it seems to slip away, which makes the meaning of intelligence and creativity feel obscure and mysterious. At any rate, creative intelligence seems to me to be a subject that is well worth exploring, especially at this time of technological revolution when the subject of artificial intelligence is on everyone's mind.

This entire book should be read as my attempt to explore creative intelligence and get a feeling for the various ways in which it arises, especially in mathematics. Let me make clear from the outset that not only do I not have an easy answer to the question, "What is creative intelligence?" or even to the questions, "What is intelligence?" or "What is creativity?" There is a good reason for this. We will soon see why such a formal definition is not possible.

When I use the term creative intelligence I am referring to one singular thing, not a hybrid of the well-defined concepts of creativity and intelligence. I believe I know what it is, but I am having difficulty writing down a succinct definition for reasons that I hope will soon become clear. Creative intelligence is something of inestimable value to every person and to society. I would even go as far as to say it is not a concept at all but something very basic to our existence, something that sustains us and the world. It cannot be described in words because it resides in a domain that is prior to words. It is, you might say, the impetus that produces words, concepts, and ideas. You can think of it as the drive towards complexity, evolution, and development in general, both in the natural world and in human beings.

All of the above is very nice and may even be a little poetic, but what is it? How would you know it if you encountered it? What is the nature of this jewel of reality? What is its place in the world? How does it find expression in the human domain in general and in mathematics and science in particular? Moreover, what can creative intelligence tell us about the contemporary world, about Artificial Intelligence and technological change?

In an attempt to explore these questions, I shall begin by saying that the expression creative intelligence is redundant because the type of intelligence it refers to is necessarily creative. Moreover, creativity, true creativity, is always intelligent. It might be useful to think of the expression "creative intelligence" as a metaphor. In reality, it is two metaphors, namely, *intelligence is creative* and *creativity is intelligent.* These two expressions are not identical; each expression carries its own distinct meaning. Like every good metaphor, these two expressions give expression to a new vista of meaning. And like every good metaphor, "getting it" calls for a creative leap. Getting it in this case means getting some feeling for what I mean by "creative intelligence."

As I said above, such ideas can only be imperfectly captured through direct definition. There is a reason why important notions like this are not amenable to a formal definition, and this reason is basic to the approach taken in this book. Let me try to explain what is going on here with an analogous but more elementary idea, the idea of "nothing." It is not as though "nothing" cannot be defined, as the Ancient Greeks believed. You can define nothing as zero, and so turn "nothing" into a fundamental and powerful mathematical concept. Nevertheless, this definition comes with a cost. The cost comes from the fact that "nothing" is obviously not "something," much less an integer in our number systems. Both the gain and the loss arise from making the paradoxical idea "nothing" into an ordinary mathematical concept like any other. The problem here is that "nothing" lives in a different world than "zero." "Nothing" lives in an implicit and informal world while "zero" lives in the explicit, formal world of mathematical structure. Exactly the same kind of problem arises when we try to define infinity or randomness or even number, time, and space.

This is also the problem that we encounter when we try to give creative intelligence an explicit definition. To repeat, this does not mean that "creative intelligence" does not exist, but that its dwelling place is not the world of concepts, ideas, and structure.

Nevertheless, we can begin to see the stakes involved in this discussion when we consider the nature and implications of Artificial Intelligence (AI). What is the "intelligence" that is referred to in Artificial Intelligence? Is its meaning here identical to the meaning of intelligence when it is used in psychology or in everyday life? Well clearly the meaning has been modified since we use the term "artificial" to refer to it. Another question is whether an AI system is, or could be, creative. These questions are extremely relevant today and, indeed, it is no exaggeration to say that they are of vital importance to our collective future. One of the aims of this book is to shed some light on questions related to the nature AI systems and the implications of their widespread adoption.

When I began writing this book, it was not my intention to emphasize AI. I wanted to write about mathematics and science from an original point of view. However, I found myself unable to avoid writing about recent technological developments. Upon reflection, it seems that the subject of AI is directly connected to the themes of this book Thus, I decided to add a chapter on AI towards the end of the book since, by that time, I would hopefully have built up an intuition for creative intelligence from various sources: child development, the structure of the brain, events from the history of math and science, the process of creating mathematics as well as the process of learning and teaching mathematics. Then we might be in a better position to ask whether AI involves creative intelligence. To be clear, I am not expecting a clear-cut answer so much as a deeper appreciation of what is involved when we take up the questions around the significance of AI. So I ask you, the reader, to bear with me as we proceed.

We often imagine that giving a name to something means that we have captured it in some way, but in the case of creative intelligence, we shall have to work very hard in order to develop some intuition for what we are referring to. The discussion is inevitably subtle since "creative intelligence" refers to something that necessarily makes contact with a domain that is prior to consciousness, and so words inevitably fail to capture it completely. My job is to try to point the reader in the right direction. One way of beginning this familiarization process is the negative one of establishing what creative intelligence is *not*.

Creative intelligence is not the intelligence of systematic thought. It is not to be found in axiomatic systems or in logic. Systematic thought is useful, even essential, in many situations, yet its proper domain comes subsequent to an act of creative intelligence. Creative intelligence is what produces systematic thought in the first place not, as many believe, the reverse.

Creative intelligence contains elements that appear opaque and mysterious to the rational mind. There is a good reason for this, and it is not something mystical. Anyone who has resolved a problem that seemed to be intractable at first, anyone who has ever thought up a new idea, is familiar with the mystery I am speaking of. It is the mystery of "Where did it (the idea) come from?" An honest answer is to admit, "I don't know." This is what I mean by saying that the origins of creative intelligence are invisible to one's conscious mind.

This same mystery is at the origins of all great art and all paradigm changing scientific breakthroughs. This mystery gives rise to the feeling of wonder that you get so powerfully in the writings of Einstein, for example, and I wrote about in my book *The Blind Spot*. The universe is so extraordinarily beautiful and complex. Is it not amazing that our minds are able to peer so deeply into the nature of the universe? That we can do so, albeit imperfectly, is a kind of miracle for which there is no logical explanation. It is a mystery. Our own minds are a mystery. Consciousness is a mystery. We are immersed in mystery and wonder yet are so often incapable of seeing it. How sad!

The entire universe is a domain characterized from its inception until today by creative intelligence. You could say, "In the beginning there was creative intelligence." Thus, creative intelligence does not only deal with the intellectual domain of thought and ideas. It is active there, certainly, but it is not there exclusively. For example, why think of evolution as driven by a haphazard and unintelligent randomness? What about the possibility that evolution consists of the manifestation of creative intelligence? I know some people will be alarmed at this idea because it might seem like bowing in the direction of religious fundamentalism. But actually, it goes in the opposite direction from formulaic ideological beliefs. A new approach to evolution requires us to tweak the meaning of randomness, which turns out to be one of those difficult mathematical ideas that need to be approached as we did the idea of "nothing."

What if creative intelligence operates equally in the natural world and in the minds of human beings? Science usually explains intelligence as the result of neuronal complexity. I would maintain that this is another of those causal relationships which need to be reversed—neuronal complexity is the result of creative intelligence.

However, creative intelligence is also at work in human beings, in the way the mind produces ideas and theories that model the natural world. When creative intelligence operates in the human domain, I shall call it deep thinking. Deep thinking, as we shall see, occurs in the creation of scientific and mathematical theories, but also in an essential kind of learning. Deep thinking, again, is the manifestation of creative intelligence.

When a scientist or a group of scientists produce an extraordinary theory, such as the theory of relativity or the recent theory of quantum fields, one is in awe of the creative intelligence involved. Nevertheless, there is inevitably something missing from the formal theory or, for that matter, from most expositions of theory. This missing element is the creative intelligence that is operating in the scientist herself, the creator of the theory. Controversial as I am sure such a declaration may be, any complete theory has objective and subjective dimensions. Objectively, there is theory and experiment; subjectively, there are the minds that conceive of the theory, integrate it with other theories, and validate it experimentally. In our search for objective certainty, we pretend that our theories stand outside the human domain and are independent of the creative intelligence that produced them. Absolute objectivity is a myth, as we shall see. We are all the creators of our personal worlds, that is, we are fields within which creative intelligence manifests itself. Moreover, when we stare up at a star-filled sky, listen to a great symphony, or attempt to grasp a new scientific insight, the awe that we feel is a manifestation of creative intelligence. It is the awe that arises as a result of the resonance between the creative intelligence of our minds and its manifestation in the universe and in scientific theory.

Deep thinking and thus creative intelligence are always present in paradigm shifts in mathematics and science. However, it is also present in less dramatic situations—in problem solving and many learning situations. Students often ask when introduced to some new idea or concept, "How did she come up with that idea? Where do these ideas come from?" The answer that is often given, unfortunately, is that the people who conceived these ideas, Einstein and Newton, for example, were geniuses and so, somehow, outside of the ordinary human domain. In fact, for many people this is the very definition of genius. However, if we look hard we shall see that deep thinking and creative intelligence are the legacy of all. This does not make the origins of genius any less mysterious, but it becomes more personal. It does not just happen to exceptional others. It has happened to you in the past and can still happen again.

Just because it is claimed that there is an element of mystery in deep thinking does not mean that we must refrain from exploring it. On the contrary, it is most important that we do probe it. The way to start this investigation might be precisely with the feeling of mystery that attends creative leaps. Why do the origins appear mysterious? It is that we are approaching things from our normal state of waking consciousness. What our consciousness cannot grasp, we consider to be mysterious.

As I said above, a direct way of probing this mystery comes from asking oneself the question, "Where do ideas come from?" or "Where do thoughts come from?" When one asks where thinking comes from, it is easy to see that such a question is difficult, even paradoxical. In one way, the question of where an idea comes from has no answer. Anyone who has had a new idea will testify that their experience was that the idea just arose (usually after a great deal of work). One day, one moment even, the idea was there in their mind whereas the previous moment it was not. The idea arises, seemingly from nowhere. What, we must ask, is the nature of this "nowhere?"

"Nowhere," in the previous sentence, as a reference to the place where idea comes from, is a place that the creative person often is reluctant to discuss, even to make conscious, for fear that making it conscious would block their creativity. "Nowhere" refers to what I shall call the implicit, the non-conscious, or that which is potentially conscious. This realm of potential can never be captured in words or symbols, since making it conscious moves it from the potential to the actual. This does not mean that what is potentially conscious does not exist. It certainly does, and without an acknowledgement that such a domain exists, the essential nature of the creative process will never be understood. It is the fertile source of all that we know and all we wish to know.

We are so used to identifying reality with the field of conscious experience that the potentially conscious may seem unreal, even mystical. And yet whenever we learn something new, get a new idea, write a sentence, or even engage in a conversation with a friend, the potentially conscious is there to be seen. When I'm engaged in a discussion with a friend or colleague, it often proceeds as follows: She says something and spontaneously I respond. I don't sit there for a time and formulate my thoughts; I respond immediately. Where was that thought of mine before it came out of my mouth? It was in the domain I am calling the potentially conscious.

The historical Buddha felt that this point was crucial, and in order to draw it to his followers attention, he said the following. Suppose you hear a sound, let's say of a gong sounding, "Bong!" Then the Buddha asked the following pregnant question, "When the sound of the gong is no longer heard, do you stop hearing?" He was referring here to the potential for hearing. The conscious experience is the "Bong" that comes and goes, starts and stops. But the potential for hearing is always there as long as you are alive. Even in a coma, it is said that the potential for hearing is active. Thus you can potentially hear, potentially see, potentially understand, and so on. Creative intelligence contains an element of this "potential awareness." Without an acknowledgement of the preconscious domain out of which scientific theory arises, our description and appreciation of science will be missing an essential ingredient.

Another common metaphor for the potentially conscious is the ocean, where consciousness would consist of the waves that make up the ocean's surface layer. The deep ocean is always there and without it, there would be no waves. It would be silly to assert that the deep ocean does not exist, yet that is what we do when we insist that reality only consists of the contents of consciousness.

A third metaphor involves the brain. If an idea or thought involves a pathway of neural connections, the potentially conscious might include all possible pathways. A simple example occurs when you cannot access something in memory, a name for example. In one sense, you know the name since if I ask you, "Is the name Fred?" You immediately answer, "No." Then a few moments or hours later, it occurs to you what that name is. It was there all the time. It was potentially conscious but for the moment inaccessible. This example shows clearly that there is much more to the mental realm than the contents of consciousness. Deep thinking necessarily requires access to the potentially conscious (which is the reason for using the word "deep").

A final metaphor that inspired me greatly comes from Quantum Mechanics and in particular from Quantum Field Theory. Quantum Mechanics is in many ways counter-intuitive, but for me, it is a profound metaphor into the nature of reality. As a recent book has it, reality is not what we naively think it is. Moreover, our own minds are much deeper and more interesting than we usually believe. Quantum Field Theory aims to unify all physical theory in one vast theoretical framework. I suggest that the metaphoric structure of this theory extends even further—to the mind itself. Just like a physical field exists everywhere and nowhere but manifests as the particles that make up the objects of our world, so also Mind is a field. It manifests itself, that is, its quanta consist of our thoughts and ideas. The mind, thinking, and creativity have an analogous structure to that of the quantum world.

Quantum Mechanics also contains the notion of complementarity, the complementarity of the wave/particle duality for example. Is the electron a particle? Well, no, not completely. Similarly, it is not just a wave. The electron can only be described completely as a wave/particle complementarity. I wrote a number of books that made the case that complementarity was a form of ambiguity and that ambiguity, as I used the term, is basic to mathematics and to science. Ambiguity here implies that there are two equally valid ways to think of an electron. When it interacts with something else, it is a particle. In its pristine state, it is a probability wave. The certainty which we associate with science only refers to the particle dimension. The wave description is necessarily uncertain. What a powerful metaphor this is for the mind of creative intelligence!

Finally, let me say that this book consists of a substantial revision of my earlier book "Deep Thinking: What Mathematics Can Teach Us About The Mind." The perspective of creative intelligence is new as are the chapters on "Mathematics as Process" and "Artificial Intelligence." The other material has been modified to reflect the changed perspective.

W. Byers
Montreal
2025

Contents

Chapter 1

Deep Thinking and Creative Intelligence

1.1 Introduction

In the preface, I discussed creative intelligence and said that deep thinking is the name I would give to the manner in which creative intelligence operates in the intellectual sphere, especially in mathematics and science. The reader might object that this merely reduces one obscure notion to another. It is to respond to this anticipated objection that I shall use this chapter to introduce what I mean by deep thinking and, in this way, get some insight into creative intelligence.

It is clear from my discussion in the preface that it is not possible to begin the discussion of deep thinking with a formal definition. The reason for this is clear—deep thinking arises from sources which are pre-consciousness. If we are to include this essential element of deep thinking, we cannot work by writing formal definitions and deducing consequences.

Instead, I shall proceed by exploring deep thinking in a variety of situations from which I hope to isolate a number of its basic properties. The first situation I shall look at involves recent work in child development, more specifically, the development of the number concept in infants. This will lead me to examine various conceptual systems for number from two points of view—the manner that they arose historically and the way in which children learn them. These sources will be enough for me to list a series of tentative characteristics of deep thinking. This will give us a pretty good feeling for deep thinking—enough, at least to continue the exploration.

However before we can describe deep thinking let's spend a moment considering the more general question, "What is thinking?" Thinking is a mental process, a way of using the mind, which involves directing it towards something—normally a sequence of ideas, statements, or propositions. Thought, which is the content of thinking, usually occurs through the medium of words but it is also possible to think using non-verbal elements such as symbols, images, sounds, or even feelings. Nevertheless for most of us thinking is predominantly verbal.

Rational thinking consists of a sequence of propositions arranged in a logical order that is organized so that each step follows from the preceding one by means of the rules of logical inference. The first statement is usually called the hypothesis and the last, the conclusion. For some people rational thinking is the highest form of thinking and all other ways of thinking are deficient.

The fact remains, however, that there are other ways of using the mind, which are neither rational nor sequential. For example, the Nobel Prize-winning economist Daniel Kahneman in his brilliant book, *Thinking, Fast and Slow*, argues for two different ways of thinking. The distinction that he makes between fast and slow thinking has a strong affinity with my distinction between rational and deep thinking. "Deep thinking" is the name that I gave to a certain kind of non-sequential thinking. Most people are not normally aware of the existence of such thinking nor that they are capable of it. Non-rational thinking would appear to be amorphous and vague precisely because we are used to identifying thought with rule-based rationality, and what is meant by rationality is a kind of sequential process such as one finds in a mathematical proof or a computer program. (I shall propose a more expansive take on rationality in Chapter 7.)

One way to get a feeling for the difference between slow or logical thinking and deep thinking would be to recall that deep thinking is a form of what I called creative intelligence, in other words, deep thinking is necessarily creative. Deep thinking is the way the mind is used in creative work. Creative thinking is not strictly rational but neither is it irrational. The relationship between rational thinking and deep thinking is important and one of the themes of this book. For now let me just say that the discovery of rational thinking was a creative act that initially arose out of deep thinking. Logical thinking has its role in the various processes that I shall describe but it is a role that is secondary to the more basic process of deep thinking.

Deep thinking is something that is difficult to get a handle on, not because it is complex, but, on the contrary, because it is so basic. Deep thinking is the way the mind functions naturally, not something that the mind must be coerced into doing. The "default condition" of the mind is active, dynamic, and creative. It does not have to be "turned on." It arrives on the scene already turned on. We shall get evidence for the dynamic default condition of the mind by considering the way infants think as they make their initial attempts to get a conceptual understanding of the world. It is from recent work on the development of conceptual systems in infants that we shall begin to discern some of the essential ingredients that make up deep thinking. For this reason yet another term for deep thinking might be "developmental thinking."

Such thinking is universal—every child is capable of it. Since such thinking is essentially creative and every step in the conceptual development of children requires an act of creativity, it follows that children are naturally creative. This means that the potential for creativity also exists, latently at least, in all adults. It is not just reserved for the brilliant and talented. Why do most adults lose touch with it? Why does our mind, the adult mind, so often feel torpid and dull? Why does it take so much effort to turn it on? Why is creativity so hard? These are all excellent questions but for now it is enough to make the observation that deep thinking is the natural condition of the mind even if this state is difficult for most people to actualize.

In other words the usual order of things is reversed. Instead of viewing creativity as a higher form of rational thought, rational thought is seen as a way of capturing and amplifying certain aspects of deep thinking. Deep thinking is more basic.

The word "deep" in deep thinking not only refers to the fact that deep thinking carries elements of the non-conscious but also refers to the property of being fundamental. Thus, it is natural to turn to the development of concepts in young children, especially one of the most elementary of all concepts—the concept of number.

1.2 Deep Thinking in the Conceptual Development of Infants

Our guide into this foray into child development will be Susan Carey, Professor of Psychology at Harvard University and her recent seminal book, *The Origin of Concepts.* Carey's book is built around three major theses. The first is that human beings (and non-human animals as well) "have systems of core cognition, and that core cognition is the developmental foundation of human conceptual understanding."[i] The existence of such core systems—systems that are built-in to our biology—suggests that the conceptual development of infants builds on a foundation of earlier biological development. There appears to be an arc of development that begins with the evolution of brain and mind structures, proceeds to the development of core conceptual systems in infants, then to the development of the more sophisticated conceptual structures that young people learn in schools and universities. This arc culminates in the creative research of science and mathematics. At every stage in this entire developmental arc, from the primitive to the most sophisticated, it is possible to discern the operation of a similar process of cognition—deep thinking.

"Carey's second major thesis is that new representational resources emerge in development—representational systems with more expressive power than those they are built from, as well as representational systems that are incommensurable with those they are built from. That is, conceptual development involves theoretically important discontinuities."[ii] Here Carey isolates a number of essential characteristics of deep thinking, which involves an indispensable element of discontinuity, that is, a jump from one point of view to another. Further the new point of view is incommensurable with the old. This implies that deep thinking is always difficult.

Carey's third major thesis is "that the bootstrapping processes that have been described in the literature on the history and philosophy of science underlie the construction of new representational resources in childhood as well."[iii] This is an attempt to explain the great conundrum of both development and creativity. If it is so difficult then how does it happen? How do development and learning come about? How do human beings manage to come up with ideas and theories that are fundamentally new and constitute a radical break with the past? Where do these new ideas and representations come from? It is all quite mysterious and even miraculous.

I find the argument around the existence of systems of core cognition fascinating but it is not my intention to go into it here. I will merely accept this thesis even though it may be controversial to some. It supports the main thrust of my argument but is not crucial to it. My interest in this chapter is in the second and third theses—the discontinuous emergence of new representational systems and the way in which they come about.

It is interesting that Carey in her third thesis uses the literature of the history and philosophy of science to support her work in child development. I also intend to use both of these sources, conceptual development as well as episodes from the history of science, to support my general thesis that there exists a common root to learning, creativity, and development, namely what I am calling deep thinking.

1.3 Core Cognitive Systems for Number

I now turn to Carey's description of a couple of systems of core cognition that deal with what has been called the "number sense."[iv] She says, "The number sense is a paradigm example of core cognition …" and lays out the evidence for two distinct systems of core cognition with numerical content.

The first of these core systems is the "analog magnitude representation of number. Number is represented by a physical magnitude" (say the length of a line segment) "that is roughly proportional to the number of individuals being enumerated." Thus the number seven would be represented by a line of length that is seven times as long as some unit length. This system is still operational in all of us. It is not precise since most people would have difficulty telling the difference between a line segment of length twelve and one of length thirteen. The fascinating thing about this core system is that it can discriminate two magnitudes as a function, not of their cardinality or their difference, but of their ratio.[v] So a six-month old infant can discriminate successfully between 24 objects and 12 objects (2 to 1) but not between 24 and 16 (3 to 2). Carey maintains that animals also represent number in this way. She produces data that supports "the existence of an evolutionary ancient representational system in which number is encoded by an analog magnitude proportional to the number of objects in the set." It is important to Carey's argument that such "number representations are conceptual; their content goes beyond spatiotemporal and sensory vocabulary."[vi]

Then Carey turns to a second core system with numerical content: "parallel individuation of small sets." This system stores in parallel, that is, simultaneously, the number of objects in small collections of 1, 2, or 3 objects. This is a precise representation but it applies only to very small collections of objects.

"There is massive evidence for two distinct core cognition systems with numerical content, in one of which (parallel individuation) number is represented only implicitly and in the other, number is represented by a mental magnitude that is proportional to the cardinal value of the set of items under consideration."[vii] [This] "... raises several questions about the relations between them. In what senses are they distinct? Do they ever become integrated in a single system for representing number? If so, when and how?"[viii]

Carey emphasizes the radical difference between the two systems. These differences include the fact that the former system number represents magnitude and is geometric, continuous and approximate whereas the latter represents multitude[ix] and is discrete and precise. Why do we have more than one core system for number? Isn't number a singular concept so why don't we have a unique way to represent number and then develop this basic notion as we grow and are educated? The fact that evolution has provided us two (or more) systems for number is an important indication of how development and learning proceeds. There wouldn't be the same need for development if not for the situation of having two separate and essential systems that carry different perspectives on number. All of the historical extensions of the idea of number—including the fractions, negative numbers, real numbers, and so on—are resolutions of the tension that is implicit in our core systems. Without that tension there might well have been no development.

In my book, *The Blind Spot*,[x] I discussed at length this tension that lies at the heart of the idea of number and the way in which this ambiguity has been resolved in many different ways in the history of mathematics. For example, the Greeks discovered that the root of two was a fine geometric number since it arose as the length of a line segment in Pythagoras' Theorem but that it was difficult to incorporate this number into the world of multitude since it is irrational or incommensurable with the rational numbers. This incompatibility between the discrete and the continuous is only really resolved in a satisfactory way with the invention of the real numbers.

To repeat, the conflict between an infant's core conceptual systems provides the impetus for development, which is fuelled by the need for some sort of resolution. The implicit conflict is resolved through the development of a new conceptual system—the counting numbers. However this resolution is not complete or definitive. It is a step on the long voyage that constitutes the development of our modern notion of number. On this voyage the primal differences between our core systems always remain active and so remain capable of generating more complex resolutions.

1.4 The Counting Numbers

It is surprising how many people still believe that numbers derive from counting—that counting is primitive. The developmental evidence that I have been summarizing tells us that the counting numbers are not at the origin of "number" but rather that counting numbers develop from more primitive core systems. Nevertheless the counting numbers (1, 2, 3, and so on) are a vital step in every person's cognitive development. Today, it is the first conceptual system for number that is learned. The system of counting numbers integrates major aspects of the core numerical systems while reconciling many of their conflicts. It supports new computations such as addition, subtraction, multiplication, and division. It is a sophisticated system, which is effectively infinite in terms of its potential for further development. Even today it is a major area of research in mathematics (number theory) with very concrete and practical applications in areas such as encryption.

This system of counting numbers is the context within which many fundamental concepts are developed. These include linear order, cardinality, and infinity, to name just three. Yet young children become quite familiar with it—familiar enough to answer the question, "What is the largest number?" with the sophisticated answer, "There is none—they go on forever." Much of elementary arithmetic is concerned with grasping this conceptual system, making it one's own. The counting numbers are the most elementary example of the way human beings, even at a relatively early age, can access and work with cultural systems of incredible subtlety and depth.

Many people's understanding of number never goes much beyond this basic system. For them a number always remains a counting number. And so it was in the history of mathematics where, as we shall see, even the recognition that the ratio between two numbers, such as the ratio of 2 to 3 is (or can be represented by) a number, is a major development that takes us into a new conceptual system.

The mathematician Leopold Kronecker famously remarked that "The integers were created by God; all else is man-made." However it is much more likely that the positive integers are learned. Carey makes a detailed and convincing argument, that I do not have the space to replicate here, to show two things. The first is that the movement from one conceptual system to another (here from the core systems to the counting numbers) is discontinuous. The second proposes a mechanism for this movement, which she calls "Quinian bootstrapping," and I shall discuss later on in this chapter.

It is the movement from the core systems for number to the counting numbers that allows Carey to isolate crucial features of deep thinking. "... If CS_2 (the more developed conceptual system—here the counting number) transcends CS_1 in the sense of containing concepts not represented in CS_1, it must be the case that CS_2 is difficult for children to learn." The discontinuous learning that is involved in the mastery of a new conceptual system is inevitably difficult. When we come to look more closely at the nature of this difficulty we shall see that learning a new conceptual system is hard in precisely the same sense that being creative is hard. In fact learning a new conceptual system is *the* quintessential act of creativity. Continuous learning, which means acquiring the technical skill to make computations on the basis of a previously acquired conceptual system, does not carry with it the same sort of difficulty. For example, learning the multiplication tables and then the algorithm for multiplying two three digit numbers takes work but it a different kind of work that is required to move from living in the conceptual universe of the counting numbers to that of fractions.

Let me end this section with another cautionary note about the danger of identifying "number" with "counting number." Even Carey slips into this mistake at various places. However there is a fundamental difference—in fact a chasm—between them. We all know what the counting numbers are but "number", in its most general and basic sense, lives neither in the world of formal mathematical structures nor in the world of cognition. It is not, strictly speaking, either an object or a concept. It is an aspect of reality that is necessarily informal and so can be represented in multiple ways by means of different number systems. Number is not so much a formal concept as it is what could be called a proto-concept, which generates multiple concepts.

Number is very subtle—you can't pin it down but you can't say it is nothing. One might say that number is a way of evoking the unity of the world as it manifests itself as order and pattern. We might identify it with "a tendency towards a complex order" that is present both in our mental processes and in the natural world. This is something the Greeks understood when they asserted that "the world is number." It is the deeper meaning of the old saying, "God is a mathematician" and possibly what the psychoanalyst Carl Jung meant when he asserted that number is an archetype. Number is informal yet incredibly significant. It is no exaggeration to say that it is perhaps the most important foundational element of our entire scientific and technological civilization.

1.5 Core Systems for Geometry

Carey's colleague Elizabeth Spelke[xi] and her co-workers Sang Ah Lee and Véronique Izard have extended Carey's work on number to geometry. They claim that an analogous situation occurs in this domain, and hypothesize: "Like natural number, natural geometry is founded on two evolutionarily ancient, early developing, and cross-culturally universal cognitive systems that capture abstract information about the shape of the surrounding world: two *core systems of geometry*. … Children … construct a new system of geometric representation that is more complete and general."

It is fascinating that Spelke, like Carey, posits two core systems. Her work follows Carey's lead in viewing development more as an outcome of the integration of the two core systems. Nevertheless it would take me too far afield to discuss geometry in its own right at this time. I mention geometry here because of the danger that artificially separating "number" and "geometry" in this way will result in a misunderstanding of both, especially the development of number. The development of the concept of number does not only arise out of the activity of counting, as one might believe from reading these researchers, but equally from the activity of measuring. For the Greeks one could argue that measuring was the more important source—numbers *were* measuring numbers, especially the length of line segments. In other words number has its origins in multiple core systems, which are antecedent to both counting and geometry.

The development of mathematical concepts does not proceed with a number concept that goes from "numerical" core concepts to counting whereas that of geometry goes from the core geometric concepts to a more sophisticated "Euclidean" system. Rather the same core systems give rise to different conceptual systems that involve different resolutions of the underlying discrepancies. Number is as much a measuring (geometric) concept as it is a counting object. Both counting numbers and measuring numbers are resolutions that succeed in resolving some but not all of the ambiguities inherent in the core systems. Both contribute to the development of more sophisticated notions of number.

1.6 The Rational Numbers as a Conceptual System

The fractions are the conceptual system that is learned after the counting numbers. Number systems in mathematics are basically situations in which you can do arithmetic, that is, you can add, subtract, multiply, and divide. However closure under these operations is sometimes incomplete. For example, adding or multiplying two counting numbers results in another one but subtracting or dividing may or may not do so. You can't subtract 8 from 3 in the counting numbers but trying to do so will ultimately lead to the creation of a new number systems—the integers—and its associated conceptual system, as we shall see in Chapter 3.

In this section I shall focus not on subtraction but on division. You can divide in the system of counting numbers. You can divide 12 beads into 3 collections and see that there are 4 beads in each collection. Alternately you can cut a pie into 12 regular pieces and consider that a sub-collection of 3 pieces bears a certain geometrical relationship to the whole pie. In other words a fraction is introduced as a relationship between two counting numbers, here 3 and 12. At this stage it is a relationship and not yet a number. You could also call it a computation over the counting numbers because the operation of division is occurring within the conceptual system of the counting numbers. You can divide whole numbers perfectly well without ever developing the concept of a fraction as we all did when we first learned that "three divides into seven two times and leaves a remainder of one."

But fractions are more than a representation system for ratios. To appreciate this extra dimension it is necessary to ask oneself in all seriousness, "Why is a ratio a number at all?" Don't answer, "Of course it is!" There is nothing obvious about it. Grasping that a ratio is a number and integrating that realization with the earlier system of counting numbers is a major intellectual accomplishment that we all made as children. It is a leap to a new world of numbers—the rational numbers.

We may appreciate this leap through an anecdote that the great mathematician, William Thurston, told of his own childhood.[xii] Thurston tells of the day when he ran excitedly up to his father and told him that he had just realized that 134 divided by 29 was a number and not just a problem in long division. His father's response was, "Of course it is." A mathematician would probably say, "That's obvious." But on the contrary it is the opposite of obvious. The father was speaking from the conceptual system of the fractions in which it is indeed obvious. But Thurston was describing the moment of insight when you transcend the system of the counting numbers and make the leap to the fractions. From this point of view it is a revelation! The two responses capture what is involved in going from one conceptual system to another and point out that what is involved in the process of deep thinking includes a discontinuous leap that gets you very excited because it gives you a vision of an entirely new world.

Let's think of this developmental problem in an entirely different way and look at it from the point of view of the ancient Greeks. For the Greeks numbers were not so much counting numbers as they were measuring numbers—lengths or areas. So asking whether something is a number comes down to asking if it can be constructed—in Euclidean geometry this construction can only use a straightedge and compass.

The geometric problem is this: given line segments of lengths ***a*** and ***b***, can you construct segments of lengths ***ab*** and ***a/b***? For the Greek geometers the construction could only use a straight-edge and compass. The required construction for the product involved the theory of ratio and proportion, more specifically, that similar triangles (triangles of the same shape and having exactly the same angles) have sides that are proportional. The idea behind the construction is based on the geometric equation, ***ab/a = b/1***, where we think of all of these numbers as lengths. Draw a triangle ***ABC*** such that the length of ***AB = b*** and the length of ***AC = 1***. Extend the side ***AC*** to ***D*** so that the length of ***AD*** is ***a***. Draw a line ***DE*** parallel to ***CB*** hitting the extended line ***AB*** at ***E***. Then the length of ***AD*** is ***ab*** because the triangles ***ABC*** and ***ADE*** are similar. (The quotient can be constructed in much the same way.)

Because this construction is possible every rational number can be represented concretely as a length, and so a ratio for the Greeks moved from being merely the relationship between two lengths to a length in its own right. But children are not experts in Euclidean geometry. It is not so obvious to them that the relationship between 5 pieces of a pie and the 12 pieces of the whole pie is a number that stands on the same footing as the numbers 5 and 12. In fact at first glance ratios and counting numbers are completely different things. For a child who defines number by the conceptual system of the counting numbers the ratio of 5 to 12 (or even of 3 to 12) is definitely *not* a number. If it is a number where does it fit into the counting numbers? Is it bigger than 1 or smaller than 1? Is it even comparable to 1? Such questions are not simple. They are hard, even "impossible" if they are looked at from the "wrong" point of view.

In what sense are such questions "hard?" The answer is that the two conceptual systems in question, the counting numbers and the fractions, are incompatible with one another. A person in the first system sees number in one way whereas a person in the second sees it in a completely different way. As Carey says, "The extension and conceptual role of the concept[xiii] of number are markedly different before and after the construction of the rationals." The child's answer to the question, "How many numbers are there between one and two?" will depend on which conceptual system the child is living in at the time. In other words, the world of number changes and grows whenever you ascend the hierarchy of numerical conceptual systems. The change from one to the other comes suddenly; it has the nature of an insight. This discontinuous aspect comes directly from the strong incompatibility, or incommensurability, of the two number worlds.

Of course when you ascend to the world of fractions you learn that it contains what is essentially a copy of the counting numbers, now considered as the fractions **n/1**. From then on we successfully manage the ambiguity of considering an integer **n** as both a counting number and (simultaneously) as a fraction. The mathematician would say that we have isomorphically embedded the counting numbers into the fractions, which means that the copying process preserves all of the operations of arithmetic. The mathematician thinks of this as a formal process and so relatively easy, and does not always bear in mind that for the learner it involves the change of conceptual systems and that is always hard.

1.7 Ambiguity

The observation that the symbolic representation of number (**2** as counting number and fraction) suppresses the context (the conceptual system in question) that gives a precise meaning to the concept of number, demonstrates that even elementary mathematics contains an essential element of ambiguity.[xiv] This sort of ambiguity is not an error but an essential part of the conceptual structure of mathematics.

In recent books I highlighted the controversial notion that ambiguity is a key element in mathematics and science.[xv] Independently but almost simultaneously the philosopher of mathematics, Emily Grosholz, published an excellent book about "productive ambiguity" in mathematics and science that approached mathematics from this same perspective.[xvi] Ambiguity involves a singular situation or idea that can be represented or understood in more than one way. An ambiguous situation is one that has multiple frames of reference and these frames of reference often contain a mutual incompatibility that cannot be avoided. Thus an ambiguous situation has a problem at its core that requires resolution. This problem may be trivial or it may be profound. If it is interesting at all; if its resolution comes from a new way of thinking about the situation then we could say that the ambiguity is productive. In other words the resolution of a productive ambiguity involves a creative act.

Now this description of ambiguity applies to the two situations that I have been describing. The first involves a child's two core conceptual systems for number. I have already discussed these systems but to review:

(1) Both systems give operational meaning to the primal ideas of number and quantity. Each is consistent in its own right.

(2) They conflict with one another, that is, there is an element of incompatibility between them.

(3) There are multiple resolutions to this extremely productive ambiguity, namely, various number systems, in particular the counting numbers and the fractions.

The second example of productive ambiguity involves the counting numbers as the base conceptual system—CS_1—giving rise to the fractions—CS_2. Here again we have two different ways of giving meaning to "number." Again the two systems are inconsistent but the inconsistency is different than in the case of the two core systems. The latter inconsistency is horizontal, one might say, whereas this one is vertical or hierarchical since CS_1 can be embedded in CS_2. Now it might be possible to consider the fractions as disjoint from the integers and so think of these two systems as lying at the same level, so to speak. This is what we do when we think of **2/3**, for example, not as a number but as a relationship between numbers. Embedding the counting numbers in the fractions requires dealing with that ambiguity and also that all of the operations of arithmetic and order may be extended consistently from CS_1 to CS_2. That is, it makes sense to say ***1/2 < 2/3*** or that the '+' in '***2 + 3***' is the "same" as the '+' in '***1/2 + 2/3***'. In this situation, the word ambiguity applies at two different levels: at the level of the elements and operations in the sets involved and at the level of the systems themselves.

The conclusion is that the development of a new conceptual system is the very prototype of a situation of productive ambiguity and therefore of a situation whose resolution requires an act of creativity. The situation of the counting numbers giving rise to the fractions is repeated countless times in a student's mathematical development.

1.8 Bootstrap Thinking

Is deep thinking as it arises in the change from one conceptual system to another continuous or discontinuous? Does learning involve discrete stages or is it the result of many incremental changes? I have been told that there are computer models that seem to support both hypotheses.[xvii]

This book will recognize the existence of both continuous and discontinuous change. The discontinuity of important events in development, learning, and scientific progress cannot be discounted nor can it be replaced by an incremental sequence of continuous changes. The reason why this must be true will follow from my discussion of conceptual systems in the next chapter. The change from one conceptual system to its successor is radical because the new way and the old way of seeing the world contain an element of incompatibility. Carey follows the lead of the philosopher of science Thomas Kuhn, and calls the two systems incommensurable, but whatever you call it, this incompatibility is the reason why conceptual change is discontinuous—it is very difficult to hold both views at the same time. It is the need to resolve the incompatibility that powers deep thinking whether it comes in the guise of development, conceptual learning, or major breakthroughs in the progress of science.

Nevertheless, this does not mean that there is no room for continuous development. In fact learning, creativity, and development have a definite continuous dimension. Kuhn initially emphasized the discontinuous nature of scientific progress—paradigm change—but later on came to modify his views and accept the element of continuous change. The fascinating question to be explored is the nature of the mechanism that accounts for discontinuous development and its relationship with continuous development.

Carey describes the mechanism behind discontinuous change as "Quinian bootstrapping." "Pulling yourself up by your own bootstraps" is impossible and this highlights the mystery behind this kind of learning. She says it is like using a ladder to get from one level to another but then throwing away the ladder when you are at the higher level. Actually, it is more like climbing up to the top rung of a ladder and then taking one more step. That last "impossible" step is the mysterious creative leap.

The child initially learns a portion of the sequence of counting numbers by rote and only later understands the full meaning of the words and the sequence. The ladder in the bootstrapping metaphor is, in this case, the activity of reciting the sequence of number words. This is a rote activity, a continuous activity that can be performed even while remaining in the more primitive conceptual system. This explains the role of rote learning and demonstrates why it is important. Moreover, it shows that the opposition between rote learning and conceptual learning is a false opposition. However, if the child remains at the rote level then what is learned is severely limited—there is no growth in the child's grasp of concepts. The objective of learning is, or should be, for the child to become fluent with the new conceptual system and integrate that new perspective with her earlier ones.

Much of continuous activity, including rote learning, can be thought of as a computation over an established conceptual system. Initially, having command of a conceptual system gives the individual a new power that at the beginning may appear to be virtually limitless. However, as the exploration of the conceptual system is extended one normally encounters problematic elements—a blockage of some sort.

The demands of schooling and society, or just the need for personal growth make it difficult or impossible to ignore the problematic elements. A new way of looking at the situation is demanded. The student is pushed from behind by the need to resolve the difficulty and, hopefully, pulled forward by the support and encouragement of parents and teachers and her own intuition that a constructive integration is possible.

At the decisive moment a new conceptual system is born in which the computation in the earlier system becomes a conceptual element of the new one. In the development of the fractions ***2/3*** becomes a number instead of remaining a relationship between two numbers. In the development of the counting numbers this is the moment when the child begins to form a conceptual system that extends the core systems and the rote computations that were the basis for the bootstrapping process. Then the child can not only count: "one, two, three, ..., seven" but enumerate a collection of seven marbles or blocks and understand that it is the same "seven" in both cases. Moreover she will understand that ***8*** not only comes after *7* in the number sequence but that ***8*** is "more" than *7*. She can also subtract ***3*** from *7*, or divide ***6*** by ***2***, and discern various other patterns within the system of numbers.

1.9 The Mystery of Deep Thinking

The discussion about the development of the number concept in children tells us a great deal about deep thinking, enough to get us started anyway. In this final section of this chapter I shall enumerate some of the characteristics of deep thinking that are implicit in this preliminary discussion. However I shall not define deep thinking, for reasons that I discussed at the beginning of the chapter. Indeed the paradox of writing about that which cannot be definitively defined lies at the heart of this entire discussion.

There is a mystery about deep thinking and I shall begin by indicating why this must be so. Suppose you meet someone on the street. They look familiar but you can't remember their name. If I say to you, "Is her name Susan?" You might answer, "No! Her name is definitely not Susan."

This implies that you *know* her name since you are sure what it is not. But this knowing cannot be accessed at that time so it remains potential rather than actual. When you say you are trying to remember the name it means that you are attempting to bring it into your consciousness. These two domains, the implicit or potentially conscious and the explicit conscious thought, both represent different modalities of one's mind. I have already discussed this essential point in the preface.

Now we are in a position to reiterate what I mean by the mystery of deep thinking. The word mystery simply indicates that the process is not completely conscious as would be the case for a logical argument, for example. The word "deep" also refers to the fact that two levels of the mind are involved in acts of deep thinking, only one of which is conscious. One way to indicate the existence of the non-conscious dimension is to ask the question, "Where does the thought or the idea come from?" Ask that question of yourself and you will see that the only answer you can give is, "I don't know." When we do creative work or are concerned with conceptual development we must delve in this "I don't know." We are called upon to speak about and investigate the realm of what could be called the non-conscious mind. This is what Carey is getting at when she talks about "boot-strapping."

For example, when a child builds up a conceptual understanding of the counting numbers on the basis of the two core cognitive systems, something truly amazing has occurred. Similarly when they move on to the system of rational numbers. Something new and profound has happened. In each case, a new mathematical world has been created.

At the level of the counting numbers, the rational numbers are not yet explicit. They emerge into conscious awareness through the resolution of a "problem" within the known. Here the problem is whether a fraction, like 2/3, is a number. The fundamental difficulty associated with deep thinking is that of giving birth to the explicit from the implicit level.

Complementarity. Quantum mechanics (QM) provides an interesting metaphor for what is going on here. In QM there is a fundamental complementarity[xviii] between wave and particle. Both describe what is going on. Both are necessary for a complete description of reality. In a similar fashion deep thinking involves a figure and ground, the explicit result seen against the implicit domain of the preconscious or the potentially conscious. Something that is implicit gives birth to something that is known.

Naturalness. The fact that deep thinking is found in the conceptual development of young children leads to the conjecture that deep thinking is the natural way the mind works. It need not be taught or learned unless it has been blocked, inhibited, or repressed. Deep thinking is usually present in learning, conceptual development, and creativity.

Difficulty. Deep thinking is always generated by something that is problematic. This problem may manifest itself as an ambiguity, an incompatibility, or even a contradiction. As a result deep thinking is always difficult.

The nature of this difficulty is one of the themes of the book. However even at this preliminary stage I note that it would appear as though the properties of "naturalness" and "difficulty" are at odds with one another. If deep thinking were natural, then one would think that it would be simple yet I claim that it is always difficult. Think of a baby learning to walk—walking is natural and yet learning to walk is hard. It takes a great deal of effort and the will to overcome many failures. All significant episodes of learning and creativity will be found to have these two components: something innate that drives one towards making the effort and some obstacle that stands in the way of one's eventual success.

Discontinuity. Deep thinking always involves an element of discontinuity—you either get it or you don't. It involves an insight, a leap to a new point of view.

Uncertainty and Incompleteness. Deep thinking is grounded in the unknown and the uncertain and so deep thinking always includes abandoning the known and the certain. Uncertainty is the way things are; certainty, on the other hand, is only to be found within an established conceptual system. Thus moving from one conceptual system to another inevitably necessitates stepping into the realm of the uncertain. It cannot come by moving from certainty to certainty in the way of a logical argument.

The insight that arises from deep thinking comes with a subjective feeling of certainty but this certainty is not absolute or permanent. All systems of thought are incomplete, as Gödel taught us, and this incompleteness can be thought of as an inevitable residue of uncertainty.

Systems always contain problematic elements in a sense that is usually not clear until the implications of the new system are sufficiently explored. Often problems can be stated in the language of the initial system but can only be resolved by creating a new system. No theory or conceptual system is universally valid. On the other hand, there is no knowledge or understanding without a conceptual system. The mind is never a blank slate. (I will expand on this point in the next chapter.)

Reframing, Ambiguity, and Reification. Deep thinking involves reframing, that is, coming to look at a given situation in an entirely new way. Reframing always implies that the situation is ambiguous—the same situation can be viewed in more than one frame. Moreover, the two points of view are initially incompatible with one another. Reification is a reframing that involves a process-object ambiguity, that is, processes at one level become objects in a higher-level system. This has also been called "encapsulation," "clumping," or "block learning."

Hierarchy and Integration. Deep thinking involves proceeding from one point of view to another. On the one hand, there is invariably an incompatibility between these two viewpoints. Nevertheless, the two viewpoints can often be integrated within the second system. The integration may be accomplished by embedding an isomorphic image of the first into the second. Thus deep thinking, especially in mathematics, is often hierarchical and moves in the direction of increasing complexity.

Relation to Systematic Thinking. Deep thinking includes, but is not restricted to, elements of thought that are systematic: that is, continuous, computational, and logical. Systematic thought in this context can be thought of as computations (or explorations) that remain within a known and fixed environment. It may clarify aspects of that environment but it will never replace that environment with another incompatible system. Nevertheless continuous thinking is often a prerequisite to deep thinking.

---

[i] Carey (2009).

[ii] *Ibid.*

[iii] *Ibid.*

[iv] Dehaene (1997).

[v] This is referred to in the literature as Weber's Law.

[vi] Carey (2009).

[vii] *Ibid.*

[viii] *Ibid.*

[ix] Magnitude and multitude were Aristotle's terms for different aspects of quantity. They roughly correspond to the continuous, for example in measuring length or area, as opposed to the discrete as in counting.

[x] Byers (2011).

[xi] Spelke *et al.* (2010).

[xii] Thurston (1990), p. 846.

[xiii] Carey (2009).

[xiv] More on this in Gray and Tall (1994).

[xv] Byers (2007 and 2011).

[xvi] Grosholz (2007).

[xvii] Psychologist Norman Segalowitz in a private discussion.

[xviii] What could also be called productive ambiguity, cf. notes xv and xvi.

Chapter 2

Conceptual Systems

2.1 Introduction

Chapter 1 used the dynamics of learning and human conceptual development to reveal some essential features of deep thinking. These features were abstracted from elementary examples of deep thinking, the child's construction of the counting numbers and fractions. Conceptual development and learning, as I discussed them, involved the replacement of one way of understanding number by another, that is, the movement from one conceptual system for number to a more sophisticated system. We represented this development symbolically as $CS_1 \rightarrow CS_2$, where the arrow stands for a creative act of deep thinking. The deep thinking that is involved in the construction, deconstruction, and reconstruction of conceptual systems such as the integers or the fractions is such a normal and seemingly inevitable development that we don't often stand back and marvel at what an extraordinary accomplishment it is.

What is it about conceptual learning that is so remarkable? In the case of number systems, it may help to recall that "number" has no objective, *a priori* meaning. It is given meaning by virtue of being placed within a conceptual system. Thus number *is* a counting number or it *is* a fraction. To go from "a number is a counting number" to "a number is a fraction" is a revolutionary act, which brings the "rational" world into existence. The fraction comes seemingly out of nowhere. It certainly does not exist in the world of the counting numbers in which it is merely the computation that is called division. The reification of this computation is the quintessential creative act, which brings something new and unexpected into the world. It is not only individual fractions that come into existence in this way but also a totally new number world with its own properties and laws as well as its own problems and challenges.

In order to place deep thinking in a more general context it will be necessary to consider the nature of conceptual systems in a more general and complete manner. This will prepare the ground for a novel way of thinking about mathematics and science as well as a way of thinking about new technologies in terms of the conceptual systems that they generate.

There are two questions to bear in mind as this discussion develops. The first concerns the implications of considering a technology, a mathematical system, or a scientific theory as a conceptual system rather than merely as an objective description of "the way things are." The second involves whether one can say something useful about the movement from one conceptual system or paradigm to another in science, mathematics, and technology. I shall begin by taking up the first question.

2.2 What is a Conceptual System?

Conceptual systems are crucial cognitive structures. The natural world with its seemingly incontestable objective solidity is a construct that depends crucially on the existence of our conceptual systems. We can only experience the world through a conceptual window.

A conceptual system is an integrated system of concepts that supports a coherent vision of some aspect of the world. A conceptual system is personal; it is a "way of seeing," that is, a "way of knowing." On the other hand many conceptual systems are learned in school and exist within a social context or a particular scientific culture. As such they have objective verifiable properties. Once we have written down the axioms of the counting numbers then Fermat's Last Theorem, which states that the equation $a^n + b^n = c^n$ has no integer solutions for $n > 2$, becomes a potential mathematical truth which could be true or false and was actually proved to be true by Andrew Wiles. The intriguing thing about conceptual systems is that they possess both a subjective and an objective dimension.

By the way, it is significant that I have begun my discussion by talking about conceptual *systems* and not about individual concepts even though a more conventional exposition would go from the elements—the concepts—to the system, which would then be conceived as an ensemble of elements. The point is that the concepts have their meaning conferred upon them by the system of which they form a part. It is the conceptual system that is crucial and thus I have made it my point of entry.

2.3 Windows on Reality

There is a famous psychological experiment, the "gorilla experiment," in which people are shown a video in which a gorilla wanders through a group of people playing a ball game. Surprisingly the gorilla passes unseen by most viewers of the video perhaps because its presence is unexpected and incongruous. The lesson to be learned from this experiment is that people do not just observe what is objectively there but that perception is a function of expectation, culture, and experience. People need a window in order to be able to perceive reality and a conceptual system provides such a window. The conceptual system is the filter that human beings use to separate what is seen from what is not—the distinction between figure and ground. This distinction is basic to perception and to consciousness itself—people are only conscious of the figure not the ground. Figure and ground are not objective phenomena, but in the cases that interest us here, are determined by the operative conceptual system.

To step back from mathematics and science for a moment, it is generally the case that the reality that many believe to be the same for all people at all times is seen through the window of language and culture. At first glance this is not really surprising. Most people would acknowledge that we live in a particular culture that is influenced by our language, among many other factors. Language and culture are windows—a Parisian and a Londoner see the world in ways that are subtly different. In a bilingual city like Montreal you can see this dynamic play out every day in the political and cultural arenas. Many of the tensions that arise are due to the different cultural lenses through which members of different cultural communities use to look at some particular situation.

Even though we might all accept the role that culture plays in constructing people's sense of what is real, nevertheless almost everyone believes that there exists an objective reality that is out there and is independent of culture. Is it not true that when we walk down the street we all see the same trees, cars, sidewalks, and houses? It is precisely the existence of this universal and objective background reality that I am asking readers to question. I am not saying that there is nothing that is real—of course there is! But we can only get in touch with the real through an intermediary, the metaphorical window of a conceptual system. Most people believe that *what* they see is independent of the fact that they see it; they imagine that it is possible to separate the content of what they see from the window through which they see it. This is the same as believing that one's knowledge is distinct from one's "knowing," the cognitive mechanisms through which that knowledge is known. This belief informs the requirement that a scientific result must be "objective"—independent of the observer. If the scientific conclusion depends on the operative scientific paradigm—the lens through which it is viewed—then the very nature of "objectivity" must be reconsidered (which I will do in Chapters 6 and 7).

I shall take the position that knowing and knowledge are intertwined so tightly that they can only be pried apart in an artificial manner and at a cost of distorting the system that is being studied. In other words, without the window we wouldn't see much of anything! For example, we tend to think that data stands on its own, that it is neutral and objective, but in my view we always *view* the data, we see it within a certain context or from a particular perspective.

Consider the trajectory of a rocket. Of course the trajectory does not depend on whether the observer is Chinese or American, or on whether the observer is in a good or bad mood. It is objective in this sense. It is quite another thing to assume that the observed trajectory does not depend on the measuring process or on the system of concepts that support measurement.

To describe the trajectory of the rocket we would need to use the concepts of time, space, number, coordinate system, velocity, and acceleration at the very least. You must first make these notions precise and quantifiable or else you have no chance of obtaining an accurate description of the motion of the rocket. Obviously the trajectory of the rocket is real but what about the system of concepts that we use to describe it?

Do the concepts of time, space, number, and so on refer to things that are real, objective, and permanent? Well it is clear from the history of science that time, for example, has meant different things to different people and continues to do so. Our scientific understanding of time has evolved and changed. In Physics there is the time of Newton, the time of Einstein, and today the time of Quantum Field Theory. Then there is the time of our biological clocks, and our psychological perception of time as described by the psychologist William James, for example. Do all of these refer to the same primordial "time?" This may well be so but the question is difficult and scientists and philosophers argue about it to this day. Within a discipline like physics, for example, we usually have a well-defined and quantifiable concept in mind when we refer to time. But this precision disappears when we look back through history or even when we change disciplines.

I am *not* saying that time is not real (although there are serious people who make that claim). All of the variations on the notion of time clearly have something in common. There is a primordial notion of time that all precise formulations refer back to. For this reason I called it a proto-concept in my book, *The Blind Spot*.[i] Time, space, number, and randomness (and many other concepts) are all such primordial ideas. Time refers to something real, something that is at the foundations of life and existence. That is why we are driven to explore it and explain it. It is real but it cannot be captured once and for all since it lives in an informal realm. We usually approach it incrementally with precise definitions that work more or less well in particular situations. In doing so we make time into something that can be worked with; we embed time into a conceptual system.

So the rocket's trajectory is real but to work with it we must look at it through a conceptual system, which is also real even though it is not totally objective. It is interesting in this regard that the word "objective" can be used in two different ways.[ii] The first is "not dependent on personal opinion or prejudice," while the second is "independent of mind." The system of concepts is objective in the first sense but not the second. The latter claim is somewhat controversial and it will set the stage for my approach to science, mathematics, and learning in Chapters 6, 7, and 8. Nothing can be seen without a system through which to view it. It follows, for example, that you cannot realistically separate "what" is learned from the cognitive processes of the learner. You can't even definitively separate theoretical content from the conceptual systems by means of which the theory is understood and made operational.

You cannot do mathematics or science without a conceptual system but such systems are not objective and permanent. They are subject to change and development. Therefore we cannot claim that the reality that we experience and work with in science is independent of the mind of the scientist. The objectivity that we claim for mathematical and scientific results is *relative objectivity*. Fermat's Last Theorem is true relative to the system of the counting numbers and Newton's Law, **F = ma**, depends on having articulated a precise notion of force, mass, and acceleration.

Consider the geometric theorem that the interior angles of a triangle sum to two right angles. This theorem is true relative to one system (Euclidean geometry) and false in another (non-Euclidean geometry). Euclidean and non-Euclidean geometry are not only different mathematical systems but are also different conceptual systems. It is easy to change from one of these axiomatic systems to the other—you just modify the parallel postulate. But conceptual systems do not change easily. The advent of non-Euclidean geometry was shocking to the mathematicians of the nineteenth century. Accepting this new geometry as legitimate involved acquiring a new way of viewing the world. Or, to put it in a more dramatic way, the world itself changed with the advent of non-Euclidean geometries.

2.4 A Conceptual Window is a Point of View

Let me give a couple of non-scientific examples that may serve to clarify this essential and, for many, counter-intuitive point—facts require a cognitive context if they are to be to be meaningful. One can think of it as involving an orientation. A conceptual system is the means by which one orients oneself to a particular situation. If one is walking in the woods then it is easy to become hopelessly lost if one does not have some way to orient oneself to one's surroundings. Similarly, one has to have a way to orient oneself in mental domains such as the number worlds that have been discussed. In cognitive situations disorientation involves the loss of coherence and meaning, and, as a consequence understanding is impossible. A student who is disoriented in some learning environment only sees raw data, which she can only deal with by means of memorization. As an aside, this means that the tendency in the modern world to see the world as data is extremely disorienting. Data means nothing on its own and cannot be understood.

Historically, one of the ways that painting has used to orient the observer involves linear perspective, a technique that was invented in the fifteenth century by artists and architects like Brunelleschi and Leon Battista Alberti and later used by the great Leonardo da Vinci in paintings like *Adoration of the Magi*. In this technique, lines in the painting converge to a point, the vanishing point, on the horizon. This makes the two-dimensional painting appear to be three-dimensional because it simulates the way in which we see—parallel railway tracks also appear to converge to a point on the horizon. Thus the painter has in mind not the objective scene (in which the railway tracks never converge) but how the observer views the scene. The painting only makes sense when the observer of the painting, whose eye makes the painting come alive, is considered along with the two-dimensional canvas (the objective painting). Yet this essential ingredient, the eye, is not present in the painting; it is only implied. In an analogous way learning does not only involve some body of objective material but also the point of view that an individual brings to the material. Just as every observer brings their history and entire being to their viewing of the picture so we bring all of our cultural and conceptual baggage to every situation of learning. Even more controversially, we could say the same thing about science and mathematics. A familiarity with the cognitive system through which the results are being viewed is the thing that makes sense of the "objective" scientific situation and makes it come to life. Anyone who has struggled to make sense of a dry and abstract scientific research paper knows that it remains relatively inaccessible unless the reader is familiar with the conceptual system that lies behind it and, even then, makes the substantial effort that is required to bring its central ideas and their implications to life.

The involvement of the observer is familiar from Quantum Mechanics where it is known as the "collapse of the wave function." But it is also evident in any scientific experiment, like those in the social sciences, where the effect of the observer needs to be taken into account. This means that a scientific experiment is not just something that can be held at arm's length and examined at a distance. Every situation is not real unless it is viewed. Another way to say this is that the complete description of a scientific experiment (or anything else) must include a point of view, a way of looking at the phenomenon, which is what a conceptual system provides. The need to include the viewpoint of the observer is completely general. In some situations, like sending a rocket to the moon, we may feel that the effect of the observer is negligible and can be ignored. However even in such situations our calculations are constrained by our theoretical perspective, which includes the mathematics that we use to model the situation, the algorithms that we use to perform our calculations, and so on. On a regular basis nature still manages to surprise us and we belatedly realize that we need to expand our point of view.

2.5 The Difficulty of Deep Thinking

With this discussion of conceptual systems in mind it is now easier to understand the nature of the difficulty I spoke about as one of the characteristics of deep thinking at the end of Chapter 1. Grasping the nature of this difficulty is the key to understanding conceptual development and, as we shall see in subsequent chapters, is also the key to understanding creativity and learning.

I said above that a conceptual system is a window that looks out at reality; reality does not stand on its own but is always "viewed." However the metaphor of the window can be misleading for it may seem to imply that there is someone looking out of the window (the mind) and that there is an external scene that is viewed (the objective world), that is, that the viewer and the viewed are the primordial elements of the situation. On the contrary, it is the viewing that is primordial whereas the viewer and the viewed are secondary, subjective and objective, dimensions of this event. In other words, we have to start with the conceptual system.

This leads to the following conundrum. Reality is necessarily viewed through a conceptual system and is inseparable from the system through which it is viewed. But reality is by definition singular—there is only one reality; there cannot be two or three. Something is either real or it is not. The notion that reality is relative or that there can be two competing and inconsistent realities is disorienting and produces untenable cognitive dissonance.

Thus, having two competing conceptual systems for the same situation is experienced as unacceptable, even impossible. In my earlier discussion of "number" I mentioned that for many adults a number remains a counting number. The counting numbers capture the reality of number for them and all of the other kinds of numbers—fractions, decimals, real and complex numbers—are less compelling. This feeling frees them from having to deal with the situation of having two competing operational definitions for number with an equal claim to truth. And yet children face such a disturbing situation whenever they are faced with learning a new number system.

Having two conflicting conceptual systems is a situation that is irritating. It may even feel threatening. And yet that is the situation that arises when one is forced to admit to the existence of two different conceptual systems that cover the same ground—two systems for number, for example. From this point of view, it is astounding that conceptual systems ever change for individuals or for society in general. One might be inclined to say that conceptual change is impossible and yet it is an empirical fact that it does happen. This is what is so extraordinary about creativity and learning. To change conceptual systems the old system, the old *world*, must lose its compelling power to define reality and be replaced by an entirely new way of seeing the world. To move from one conceptual world to another, we must inevitably confront the moment when the first world breaks down and yet the new one is not firmly established. We must face this void and keep on going. This is the nature of the extraordinary difficulty that inevitably arises in situations of conceptual change.

This discussion also explains why conceptual change is discontinuous. Reality is either defined by CS_1 or CS_2 and this "or" is the exclusive "or" which does not admit the possibility that both systems are operative at the same time. Of necessity one must jump from CS_1 to CS_2. In order to capture the discontinuous aspect of paradigm change, Thomas Kuhn used the metaphor of the Gestalt picture, for example, the one of the young woman and the old lady, as I did in my books, *How Mathematicians Think* and *The Blind Spot*. Visually one can only see one of these images at one time. You either see the young woman or the old lady and the image you see is an interpretation of the entire visual field. Thus these Gestalt pictures capture the discontinuity of conceptual change very well.

2.6 Technological Conceptual Systems

Conceptual systems are the means through which the mind enters into our description of nature. We normally think of the world as being made up of things like atoms, molecules, and cells, and forget that these are not "pure" objects; they are conceptual objects. Thus the atomic theory is a conceptual system, a way of looking at the world. It may initially seem strange to say that the atomic theory of matter not only is a description of the objective properties of matter but is also a way of thinking about matter. But that is the way it is. The world that we experience and think about necessarily involves the mind.

This may seem to be a very abstract and philosophical point but in the contemporary world such questions are vitally important because of their immediate relevance to the most important phenomenon of our time—technological change. Technologies, just like scientific theories, carry conceptual systems with them. Technologies—the telephone, radio, television, computers, cell phones, and so on—are not just pieces of hardware. Their importance stems from the fact that they generate "technological conceptual systems" (TCS). Every significant new technology, the coming revolution involving Artificial Intelligence comes to mind, ushers in a new TCS, a new way of experiencing and interacting with the world. Marshall McLuhan's famous aphorism was, "The medium is the message." The medium here is the technology but the message is the TCS.

In a very real sense that is a matter of the life experience of everyone, the world changes, society changes, and individuals change with the advent of these new technologies. This explains why they are so exciting but also why they may be frightening. Young people have less resistance to change and so they are usually the first to embrace the new technology. Their parents are then shocked to discover that their kids are living in a different world than they are.

It is practically a cliché to say that we are living in a world of continuous change. In the past cultures might be stable for centuries. A change in the dominant TCS took a very long time to be assimilated into culture. Yet today all of us have experienced multiple technological revolutions in our lifetimes. The major challenge of the contemporary world involves the reaction of cultures, societies, and individuals to the challenge, not of one or two changes, but of accelerating and continuous change.

The magnitude and radical nature of what is demanded of us is clear from my earlier discussion of conceptual change. A TCS, like any other conceptual system, defines what is real for the individuals and societies that embrace it. The telephone, for example, brought about a revolution in communications, commerce and industry, not to speak of people's notion of privacy. Many people initially resisted the specter of having a telephone in their homes because they were repelled by the idea of being so freely accessible to the world, which seemed to them to be an unacceptable intrusion into privacy and the home. The telephone so completely changed people's relationship with one another that it is difficult to imagine what the pre-telephone world was like. The telephone produced a world that was radically different from the world that preceded it. And since then change has only accelerated. Today even people who climb Mount Everest carry satellite phones. Most of us may have some nostalgia for the world without instant communications but probably would find living in such a world, the pre-telephone world, to be difficult, maybe even anxiety provoking. To test whether this is so, try taking a technology fast for a few days—turn off all your electronic gadgets and monitor your reactions. One feels cut off, oddly diminished, and alone.

If a new technology changes the individual's sense of what is real then, of course, technological change is stressful and difficult. If a single change from TCS_1 to TCS_2 is difficult it pales in comparison to what is demanded of people today in all advanced societies—accommodating to repeated and accelerating change. Conceptual change is intrinsically difficult for reasons that I have discussed earlier but the change from one number system to another is clearly not in the same league as a major change in the means of communication where one's relationship with one's family, friends, and society is altered.

One basic element of human nature involves the need for security and the resulting resistance to change. As we shall see in Chapter 4, this tendency is even hard-wired into our brains. This means that once we have grasped a CS we have a strong predilection to remain there. Now conceptual systems are elastic and potentially infinite so that staying within a CS does not mean that we are doomed to go on repeating ourselves in a kind of endless loop. We can spend a good deal of time—possibly our whole lives—living within a given conceptual system, exploring the universe that it brings to life, and feeling good about the knowledge that we have acquired. In such a situation, we can feel productive and comfortable at the same time. We may not know the answer to every question but we have learned how to look for answers.

To be repeatedly forced out of our security zone is an unprecedented event in the evolution of civilizations. If the need for security and stability is indeed basic to human nature, then the modern technological world is demanding nothing less than a new and different type of human being—one who can embrace change. There have always been individuals who seem to thrive on change. But such people have been in the minority; they comprise a small number of unusual people who somehow are comfortable with change and the insecurity it brings in its wake. These people even embrace change; they find it stimulating. It makes them feel more alive! We all get some kind of inkling of this when we travel and are stimulated by the novelty of the sights, sounds, and smells that we encounter on a vacation, for example. At such times we are not in our right minds, that is, not in our habitual state of consciousness. We inhabit a different mind, so to speak, and I shall discuss this mind in Chapter 4.

Thus the current situation that we find ourselves in carries not only great dangers and difficulties but also great opportunities. Human beings do not only need security they also need to develop—to grow and learn. After all change—call it flux, chaos, or impermanence—is the way things are. Human nature at its most basic level is no different than the rest of reality in that it is continually changing. With the advent of language and science we reified change and named it. We also reified ourselves by inventing the word "I." It was only at this stage of human evolution that people came to see themselves as fixed, unchanging objects in a world of other objects. They forgot or repressed the more basic level of impermanence.[iii] However the modern world threatens to make that strategy obsolete. We look around us and find that everything is changing so fast that it is difficult to keep up. We are on this technological treadmill and the faster we run the faster the treadmill moves beneath our feet. We keep reaching for a new equilibrium because that is what always worked in the past but any equilibrium we reach gets disrupted within a short time. Equilibrium and stasis as a strategy will not work any more.

We need a fundamental shift—a new way of being which will arise by accepting change as our primary reality and learning to navigate that kind of world. This shift is equivalent to accepting that all citizens of a modern, technological society—especially those people who hope to come up with innovative ideas—must become comfortable with deep thinking. This is what I meant when I said earlier that deep thinking could no longer be ignored or repressed. Creativity is not only the birthright of everyone; it is also necessary for everyone. It is the only way to live in this peculiar world that has come into being alongside our endless stream of technological conceptual systems.

However there is the danger that we shall collectively fall into the trap of thinking that a computer can do the deep thinking for us, which means that we can remain secure without having to confront what Alan

Watts called the "wisdom of insecurity." For now it suffices to reiterate that the discussion of deep thinking and conceptual change is directly relevant to the most vital questions that can be asked about the modern world. These involve the nature of thinking and learning and so of maintaining our humanity in this new world of intelligent machines.

2.7 Conceptual Mathematics and Science

Mathematics is best understood in conceptual terms. The late mathematician and philosopher Reuben Hersh, a man who urged mathematicians and philosophers of mathematics to take another look at the nature of mathematics, commented on this in a recent paper.[iv] He first notes the reality of mathematics, which, he says, for mathematicians is as incontestable as the physical world is for the scientist. Then he asks, "What is this reality? It is our internal concepts, our mental models, which are *real objects with real properties*, (italics in original) and which are congruent to each other, which fit together and match." In other words, the reality of mathematics is a conceptual reality. Mathematics is a vast conceptual system with each sub-discipline of mathematics, such as algebra, geometry, or analysis, constituting a conceptual system in its own right. This view of mathematics is one that I shall pursue in some detail in this book. It is not the usual view of mathematics and many people would take exception to looking at mathematics in this way.

Speaking of conceptual mathematics brings to the fore the complex nature of mathematical reality with its mixture of subjective mental models and solid and objective structure. It is this complexity that we must accept and investigate if we wish to plumb the deeper nature of mathematics.

Viewing mathematics in this way will teach us what an extraordinary thing a conceptual system is—even a conceptual system that is as elementary as the counting numbers. Conceptual systems bring life and coherence to the world of our experience. The creation of a conceptual system by an individual is a remarkable event. All conceptual systems give meaning to an aspect of reality and in the case of the counting numbers that reality is the world of "number." Number is the most basic idea in mathematics and science. For those with sensitivity to mathematics, the world of the counting numbers is a vision of beauty, with its balance between infinitely nuanced intricacy and solid, enduring structure.

When the cognitive aspects of a conceptual system like the counting numbers is ignored, it can become a dry and abstract system whose essence may appear to lie in a set of axioms or the rote enumeration of the integer sequence. But when it is regarded as a conceptual system and explored from the inside, when it is grasped, if only to a limited extent, then one finds that it is teeming with life. It contains an unlimited number of questions that can be asked, computations that can be made, regularities that can be discerned, and theorems that can be proved. To discuss the counting numbers as a conceptual system is to highlight the life within the system. A conceptual system is best appreciated from the inside.

Even though a rich conceptual system like that of the counting numbers opens up a seemingly infinite world, nevertheless every conceptual system is limited in a fundamental way. Because it makes concepts and ideas (like 'number') explicit by formulating them (in a formal or informal manner), certain rigidities and other problematic elements are inevitably introduced. Within the conceptual system of the counting numbers a number *is* a counting number so you cannot subtract five from three or divide nine by four in a way that is entirely satisfactory. Every conceptual system inevitably contains a problem or limitation and, as we shall see, this can act as the impetus for the development of increasingly complex systems—in our example this would be the fractions or the (positive and negative) integers.

Science too consists of a series of conceptual systems or paradigms. Actually all of science is one paradigmatic system characterized by features such as the scientific method, logical coherence, and independence of the observer. It is easier to make the case for the conceptual foundations of mathematics than of science because in science we believe that what we are studying lives in an objective reality which we call the natural world. But science, no less than mathematics, does not work without a technical conceptual language.

Every mathematical and scientific conceptual system can be used to make computations.[v] In fact, this is one of the main objectives of such systems—to reduce investigations to routine computations. Once you have Newton's laws of motion you can perform the calculations that enable you to send a rocket to the moon. So even though the system can be finitely presented (as set of axioms or principles) it can support a potentially infinite set of computations. There are even people who believe in the possibility of an "ultimate" scientific theory, that is, a finite conceptual system that will describe all physical processes. In such a system, all phenomena will be produced by a series of computations arising from the elements of the system.

We cannot do mathematics or science without such a conceptual system, which in science is more likely to be called a paradigm. Actually there are (at least) two distinct modes of "doing" mathematics and science. The first involves working within a fixed paradigm and the second involves replacing one paradigm by another. The dominant mode of thinking in the former situation consists of continuous, analytic thought. The latter involves deep thinking but we can get some information about deep thinking by investigating the distinction between the two. What does the existence of a paradigm do for us? A paradigm is a way of interacting with the world. We can only "see" through the lens of a paradigm. It brings a particular viewpoint of the world into existence but simultaneously limits what we can see since it necessarily excludes other viewpoints. It inevitably highlights certain aspects of reality while hiding others. It is like looking through rose-colored glasses. The hue comes from the glasses and not from the scene. But even if we acknowledge this and attempt to account for it, there remain intrinsic limitations to our ability to describe the "objective" situation, which is an abstraction that does not really exist. The real world is intimately tied to, and ultimately is inseparable from, the lens through which it is viewed.

The indispensability of conceptual systems is of deep importance, a fact that is not always appreciated. For one thing, since conceptual systems are not totally objective neither are the scientific or mathematical theories that they support. Conceptual systems by their very nature are subject to change and development. It follows that neither mathematics nor science deals in permanent and unchanging truth.

We cannot claim that the reality that we experience and work with in science is independent of the mind of the scientist. The objectivity that we claim for mathematical and scientific results is a relative objectivity. The incontestable reality of the mathematical world consists of our internal concepts and mental models that are made objective by a complex social process. Having established such a system it becomes possible to say that certain propositions are true or false *relative to this system*.

The deep thinking that goes by the name of paradigm change is an extremely basic process. We have difficulty understanding such changes because we look at them from the standpoint of CS_2, the more refined system of thought, and erroneously imagine that reason can "capture" the revolutionary thought that produced the change. Such "explanations" are after-the-fact rationalizations of change. To think properly about such conceptual changes, we must place ourselves in the more elementary system from which vantage point the change is radical and unthinkable.

2.8 Deep Thinking, Conceptual Change, and Technology

As I did in Chapter 1, I shall conclude this chapter with a summary of some of the properties of deep thinking that arise from a consideration of conceptual systems. Again, I remind the reader that deep thinking necessarily contains an element that is not conscious and, as a result, cannot be reduced to words. Well, you might counter, isn't that what you are doing here? In my view, I am going as far as words will allow me to and, as for the rest, all I can hope to do is point in the right direction in the hope that some discerning people will be able to distinguish the moon from the finger that points at the moon. Deep thinking is, metaphorically, a diamond and I will be discussing certain facets of the diamond. The diamond as a whole is intrinsically mysterious and ineffable.

Remember the metaphor of the quantum particle. Just as the particle has no precise location (or any other properties) in its wave manifestation so deep thinking cannot be pinned down completely. The wave description does not allow for grater precision than this because until it interacts with something it is literally nowhere, that is, at no specific location in space and time. In fact, space and time themselves may well have such a quantum nature.

One key to this paradoxical situation is the statement, "*deep thinking comes from nowhere*." The results of deep thinking, the insight, arises in the mind suddenly but one just cannot say from where. To put this in another way, deep thinking results in a particular conscious thought or idea. It thus appears in consciousness but it does not come from a previous act of consciousness. On the contrary, it arises from a domain, analogous to the wave description, of the preconscious or the potentially conscious. Where does an idea come from? You cannot say. All you can say is that it comes to you, it is revealed, or it is manifest in your consciousness.

Paradigm change or the transformation of one substantial conceptual system into another requires deep thinking. You cannot expect to write down a prescription for such change nor expect it to emerge from some mechanical process.

Nevertheless, as we have seen, technological conceptual systems often create a new world that was inconceivable before it came into existence. This is especially true for the radical changes we can expect to accompany the introduction of Artificial Intelligence which is so important that I shall devote a chapter to it later in the book. For all such technological changes, the crucial thing is surely that we should approach technological change as conceptual change—the movement from one conceptual system to another. Like any other creative endeavor, this involves deep thinking. The result of deep thinking is not only a piece of hardware or software but a systematic way of looking at the world. Again, the essential thing is to distinguish carefully between the process—deep thinking—and the technology that arises from the process. With this comment in mind we can repeat the list of characteristics of deep thinking and see that they apply to examples of both conceptual and technological change.

Deep thinking is **natural** and as a consequence creativity reflects something basic in the human condition. The excitement that often accompanies work in modern technology is a reflection that the process of innovation touches this creative essence of the human spirit. The sense of disorientation that characterizes the early stages of adaptation to a new paradigm may be viewed as an opportunity to break through our identification with a particular culture and embrace a new one.

Changing paradigms is necessarily **discontinuous** and **difficult** both for the individual and the culture. It is radical in that it forces people to deal with the world in a different way, that is, it changes the world of experience. When we introduce new technologies into schools and other social institutions, for example, we must be aware of this, prepare for it, and introduce it in an appropriate way. One is looking for a sudden insight—a reframing. It arrives suddenly and not incrementally so the problem of introducing new technology has much in common with the problem of teaching and learning new conceptual systems that I shall discuss in subsequent chapters.

Uncertainty refers to the process of innovation. It is the uncertainty of going back to the drawing board and thinking things through with no preconceptions or, at least, as few as possible. "What if we thought about it this way?" is the kind of generative question that one might ask.

Incompleteness refers to the end result, which never accomplishes perfectly what you hoped it would accomplish. There are always glitches in any process but more fundamentally one must always bear in mind what the conceptual system can do and what its limitations are—and there are always limitations. This needs to be born in mind when one begins to interact with an "intelligent" computer system, for example. It is also something to be considered before we all start adopting driverless automobiles.

There is always a basic **ambiguity** in any situation involving conceptual or paradigm change. When one considers ideas that are basic to life such as creativity and intelligence one must remember that viewing them through the lens of a particular conceptual system, whether it be neurobiology or Artificial Intelligence, may be extraordinarily creative and produce all kinds of insight, but they will necessarily miss something. The fundamental problems of life and death, creativity, and intelligence are human problems and thus are open-ended and have no definitive theoretical answer.

Conceptual systems are also **hierarchical**—they build upon one another—and so change looks *continuous* when looking backwards from the present situation. But we have only to think of any number of instances of innovative thinking to realize that breakthroughs always come as a surprise. The **integration** that gives the impression of continuous development happens after the fact. After we have the new system or technology we think about all the cool things we can do with it and all the things that we now do in a different way that we could do better in the new system.

Systematic thinking happens within the system or technology so such systems spawn new ways of doing things. There is a qualitative difference between work that is done in tweaking a new piece of software so that it works in a new environment and creating the software in the first place. Most problem solving involves applying known techniques to new situations. A truly innovative solution involves reframing the situation. One must ask oneself what kind of problem solving is required.

Finally, there is a paradox here that is basic to what is going on. Technology is always attempting to replicate aspects of intelligence and deep thinking. Of course this could be said of any conceptual system. And this attempt is always doomed to fail in an absolute sense even though some of the attempts are extraordinary. This is something we should understand when we think about AI and think about whether we should give legal rights, for example, to AI systems. As Noah Harari says repeatedly in his fascinating book, *Nexus*,[vi] "an information system is intelligent but not conscious." When we are amazed by some brilliant technological innovation it is invariably because of the uncanny way it replicates and may even seem to exceed certain human activities or mental processes. But closed systems can never capture the fundamental openness of deep thinking for the simple reason that once it is reified it is constrained.

---

[i] Byers (2011), p. 107.

[ii] As has been pointed out to me by Albert Low.

[iii] It is interesting that in Quantum Field Theory a kind of quantum flux emerges as the ground of everything, including time and space.

[iv] Hersh (2014).

[v] I shall give many concrete examples of conceptual systems in science throughout the book—the "counting numbers" or the fractions are two of the most elementary conceptual systems.

[vi] *Nexus*, Harari (2024), p. 201.

Chapter 3

Conceptual Systems in Mathematics and Science

3.1 Introduction

Chapters 1 and 2 made the point that creative thinking is present whenever one conceptual system is replaced by another—in child development, in the elaboration of "number," in science, and in the development of technology. Accordingly it is appropriate to reexamine the nature of these disciplines highlighting the type of creative thinking that we have been discussing.

In this chapter, I consider the analogy, which is already explicit in Carey and her colleagues, between the historical development of mathematics and science, on the one hand, and development and learning, on the other. I have already discussed what it means to consider number systems such as the integers and the rational numbers as conceptual systems. In this chapter I go on to discuss other famous episodes from the history of mathematics and science. Of course there is nothing new about the mathematics and the science I shall discuss but there is something unusual in the perspective that I take.

This perspective was sketched out in Chapters 1 and 2. Paradigm change necessarily involves a discontinuous jump. Reality is singular and each paradigm evokes its own reality. This is the reason that scientific paradigms are not changed without a great deal of conflict; the reason why deep thinking is so difficult and involves overcoming so much resistance both in the individual and in the larger culture. In fact it has been said that a scientist never really gives up the paradigm within which she has been trained. What happens is that a new generation grows up within a new paradigm and the old generation retires or dies off.[i] For the most part the researcher spends her time making continuous computations within an unchanging conceptual system.

A conceptual system is inevitably associated with a particular way of thinking. Mathematics and science involve different modes of thinking, of which deep thinking is the most creative, and therefore the most difficult and the most important. A major topic in this book is these different modes of thinking and the relationship between them. I have already alluded to this in the previous chapters when I discussed the respective roles of continuous, computational thinking and discontinuous, deep thinking. At the end of the chapter, we shall see that this distinction is crucial for tackling an important problem, namely, the proper role of logical inference.

Logical deductive thinking is appropriate in some situations but not in others. The way one uses the mind within a conceptual system is not the same as the way one uses it when one makes the leap from one system to another. The former is logical in the classical sense whereas the latter is strongly non-logical. This is a huge change for the way we think of science and mathematics. It makes ordinary logic relative—part of the language that we use within a system—not an absolute truth that stands outside all systems.

3.2 Heliocentric or Geocentric?

I begin with the question of whether the earth or the sun is at the center of the cosmos because here my comment that a conceptual system creates a world can almost be taken literally. All human cultures have a mythology that describes their place within the cosmos. Of particular interest, as we shall see, is the geographical *center* of that cosmos, for the location of the center provides an orientation for the entire culture. Here we discuss three possibilities: (1) geocentric—the earth at the center; (2) heliocentric—the sun at the center; and (3) no geographical center. We might ask, "Which one of these possibilities is true?" But when we think of them as three different conceptual systems we should rather ask, "What are the properties, advantages, and disadvantages of each world view?" We don't ask which is true and which is false so much as we ask about the implications of subscribing to one view or another. In fact, these three conceptual systems constitute major stages in the development of modern culture.

Today we would say that neither the earth nor the sun is at the center of the cosmos—the cosmos has no center. In this sense both the geocentric and the heliocentric theories are false. But in another sense neither is false. When one makes a mathematical model of certain situations it is often necessary to set up a system of coordinates, normally the mutually perpendicular x-, y-, and z-axes. The intersection of the three axes is the origin. In a sense the choice of the origin is arbitrary—any point would do. You pick a particular origin so as to simplify the resulting equations as much as possible. From this point of view whether you pick the earth or the sun to be the centre of the solar system is not important. Either one will give you a coordinate system relative to which you can describe the motion of the planets. This is not to say that there is no difference between the two systems. In the geocentric system the earth is fixed but the orbits of the planets are quite complicated. In the heliocentric system all of the planets, including the earth, stand on an equal footing. Their orbits are all ellipses, one of whose foci is the sun. This facilitates computations, which are notoriously difficult in the old geocentric system of Ptolemy. It also makes it easier to conceptualize the solar system as one entity, to think of it as one system.

What, then, is lost by going from the geocentric to the heliocentric? What is lost is the ancient idea that "man is the measure of all things" (Protagoras 480–411 BCE). It is the beginning of a change in Western culture in which man moves from being created in the image of God, and so residing at the centre of creation, to a position which is further towards the periphery. The geocentric system not only situates human beings closer to the center of the universe but is also closer to everyday human experience. We see the sun rise in the east and set in the west. This model of the cosmos is then more consistent with the evidence of the senses.

You could say that there are two possible ways of viewing our relationship to the world. The first is that human beings are at the center of the universe and everything else gets its meaning and relevance from its relation to us. The second is that we are peripheral and insignificant. The movement from being earth-centered to being sun-centered is a movement away from the human being as the center of creation. But it is an intermediate position because the modern view is that there is nothing special about human beings, the earth, the sun, or even the particular galaxy that we find ourselves in. We have moved from a religious system, which emphasized the uniqueness of human beings, to the scientific system where there is nothing special about mankind.

Now it may well be that the "center" does not have a geographical position at all but this does not mean that it does not exist. "Centering" is a very general tendency that one can discern in a vast variety of circumstances. We can think of it as the need for an orientation, which is what a conceptual system provides. This is another reason why conceptual systems are so basic. Human beings' need for understanding can be seen in the need to be oriented. When you do not understand a number system, for example, it seems to be made up of arbitrary facts and incomprehensible rules that have no inner logic and must be mastered by rote. When you do understand a mathematical system then it all fits together in an inevitable way—it all makes perfect sense and facts that you do not recall can be worked out on the spot. You now possess a strong sense of being well oriented within this conceptual universe.

This need for a system of orientation is a fundamental human need. As a result all human societies have developed ways of "centering" themselves and symbols that represent the center. The philosopher of religion, Mircea Eliade, points this out quite eloquently.[ii] This human need for a center explains the social and cultural role of the President of the United States, of sports idols and movie stars. It explains people's attachment to symbols like the national flag or anthem. For many people the need for an external, physical centre has dissipated but if it is indeed a basic need then where has it gone? What has modernity done with the need for a center? To answer this question properly would take more space than is available to me here. I will restrict myself to a comment by the writer Albert Low who says, "In the past human beings have used many different ways of ensuring an orientation point or center, …. Originally the point of focus was "outside": a sacred tree, a totem pole, a flag, a rock, an idol, each of these has in the past served as a center, a basic orientation point around which to structure what we know as experience and existence. Nowadays we have introjected the center and call it 'I'."[iii] We often hear talk about the desirability of "being centered," and this can mean nothing other than finding a center within ourselves—in the self and personality that we construct over a lifetime. Thus the need for a personal center continues to exist but modern scientific culture sees no need to locate this centre geographically.

Be that as it may we can see that the conceptual system through which we view the cosmos has profound implications for individuals and cultures. Geocentric, heliocentric, or without a physical center represent three profoundly different ways of living in the world. The loss of an external center, which comes about with the dominance of the scientific worldview, may well have contributed to the alienation, anxiety, and general angst that are so common in the modern world.

3.3 Euclidean and Non-Euclidean Geometry

Euclidean geometry is hugely important in the cultural history of mankind. It is sometimes presented as merely an axiomatic, deductive system and, indeed, it is of great importance that a body of knowledge could be organized in this way. However this is only part of something that is even more important and basic, namely, that Euclidean geometry is a conceptual system. My earlier comment that a conceptual system creates a world is perfectly illustrated by this. Euclidean geometry is not a model of space—that is a modern way of looking at it—Euclidean geometry *defines* space. It represents our experience of two and three-dimensional space. Now this statement is not precisely true because Euclidean lines have no width and are extendable to infinity so it would be more accurate to say that Euclidean geometry is an idealization of our experience of space. For generations, people believed in this idealization, believed that it contained a truth that was absolute and objective. In fact this is one of the origins of the belief that mathematical systems accurately capture the real world, even that reality itself is fundamentally mathematical. From my point of view, this is just what large conceptual systems do—they define what is real for the cultures that subscribe to them.

I shall not repeat what I have written elsewhere[iv] about Euclidean geometry and the non-Euclidean revolution. I shall just summarize quickly what we can learn from this episode in intellectual history. The acceptance of non-Euclidean geometries as legitimate mathematical theories on a par with Euclidean geometry involved the movement from one conceptual system to another. This movement involved two radically different ways of viewing the world. The first of these worlds, CS_1, is obviously the Euclidean one but the second, CS_2, is not the non-Euclidean as one might expect. Rather it is a world consisting of both Euclidean and non-Euclidean models for space.

The transition from CS_1 to CS_2 is always difficult and discontinuous. In this case, it took thousands of years to make this transition because the Euclidean world was so deeply entrenched in Western culture. The difficulty of the transition is illustrated by the fact that Gauss, the preeminent mathematician of his day, anticipated the non-Euclidean revolution but did not publish his findings for fear of the controversy that he was sure would follow. He was right, the results were indeed controversial but he did not sufficiently appreciate how ripe the world was for this change of viewpoint.

Major shifts in conceptual systems like this one always carry with them huge consequences that could not have been anticipated by the mathematicians who supposed that they were merely questioning the validity of a particular axiom or even creating a new geometrical theory. In a very real sense this revolution created the world of modern mathematics—the world of formalism. An important aspect of this world, which now came to the fore, was the identification of mathematics with its logical structure. Not only does geometry cease being identified with space and become instead a model of space but also all mathematical theories become potential models of physical situations. Even the need for potential applications gets attenuated so that now there arises the possibility of mathematics for mathematics' sake—a new kind of "pure" mathematics. There is a shift in the ontological status of mathematics in which reality gets pushed further back. It follows that there is a radical change in the very meaning of truth, which is discussed for example in the final chapter of the book by Trudeau on *The Non-Euclidean Revolution*.[v]

This is more evidence that the development of conceptual systems is hierarchical, as we saw was true of number systems, and not merely the shift from one theory to another unconnected theory. This is important evidence for my thesis that you cannot simultaneously hold to two such conceptual systems. As I said above in the first system space is Euclidean; in the second different geometries are models of space that are equally valid as mathematical theories and may all have valid applications to certain spatial situations. Today when we look back we think that all that happened was the invention of another geometrical system by tweaking the parallel postulate. We often do not appreciate how revolutionary the change was because we are viewing the matter from the perspective of the conceptual system, CS_2. From the perspective of CS_1—Euclidean geometry in a Euclidean world—the "relative" world where any arbitrary set of axioms produces legitimate mathematics is simply inconceivable. If one does not grasp the incompatibility of CS_1 and CS_2 then one misses the essence of the situation.

3.4 Number Systems

In Chapter 1 I wrote about the concept of number. I discussed two core systems for number and two learned systems—the counting numbers and fractions. I viewed the counting numbers and fractions as conceptual systems and was especially interested in abstracting certain basic properties of deep thinking from the way in which one of these systems developed from an earlier one. Now I shall expand the repertoire of conceptual systems for number and see if the same "deep" properties can be found in these new systems.

3.5 Integers

In the first chapter I discussed the transition from counting numbers to fractions. Fractions are a natural development of the counting numbers but the development of number can also proceed in another direction and that is from the counting numbers to the integers through the introduction of the negative integers: *−1*, *−2*, *−3*, ... The (positive and negative) integers form a conceptual system in their own right. "Is division a number?" was the question that got us from the counting numbers to fractions. Here the analogous question is, "Is subtraction a number?" Again this is not at all obvious. Subtracting *3*, for example, is a computation that can be done within the system of counting numbers without the need of making **−3** into a number in its own right. Later on, when a light goes on and you see that **−3** is actually a number, you are in a new conceptual system. When this happens **−3** goes from being just a computation to having an ambiguous status (as process and object). At this stage it becomes one mathematical entity which can be viewed alternatively as a computation over the counting numbers or else as a new variety of number, a new conceptual object, that lives within another conceptual system.

The world of integers is qualitatively different than the world of counting numbers and this can be demonstrated by considering the process of multiplication. Everyone knows that **2 × 3 = 6**. If asked why, most people would likely point to multiplication as area or as repeated addition. However, most children and many adults have a problem with the product of negative numbers, such as **−2 × −3 = +6**. They accept it but it makes them a little uncomfortable; they don't really understand it. The reason for this is that they remain to a certain extent within the conceptual system of the counting numbers and the computations that are performed within it. Because the negative numbers seem to be very similar to the counting numbers we tend to handle them in the old way, with the addition of a few new rules, such as "minus times minus equals plus," which are learned by rote. Even mathematics teachers often forget that the integers are a new conceptual system for the child and that the transition from one system to another is necessarily hard. Memorizing "rules" seems to be attractive to the extent that it obscures the difficulty. It has its place, but certainly does not replace conceptual change. Eventually one has to deal with things like the problematic multiplication of negative integers.

Another problem arises from the manner in which the counting numbers are embedded into the integers. **3**, for example, is identified with itself (we could make matter clearer by saying that the counting number **3** is identified with the integer **+3**, but normally the + sign is suppressed). As a consequence when we speak of the number **3** it is not clear what system we are in. The notation is ambiguous. The appropriate context—counting number or integer—must be inferred. Fractions have the identical ambiguity because even though we may speak of identifying **3** with **3/1**, in practice we use the notation **3** to refer both to the counting number and the fraction. This ambiguity is a strength and not a weakness of the conceptual system of integers and we can expect to see such ambiguities arise whenever we move from CS_1 to CS_2. When we look at matters from the perspective of CS_1 then the problematic elements should also be regarded as an opportunity for making a necessary conceptual leap. If viewed constructively (and this is the role of good teaching) they are potential gateways to a new and higher order understanding of number. From the perspective of CS_2 a counting number is just a specific variety of integer. Since the integers form a well-defined and logically coherent system, the counting numbers are perfectly integrated into this new hierarchical structure.

Nevertheless from the perspective of CS_1 the two systems are incompatible. This is easy to see historically. Why, after all, were they called "negative" numbers? Clearly it is because there is something problematic about them. They are hard to conceptualize, because, from the perspective of the counting numbers **−3** is not a number at all. This is, and will always remain, a conceptual problem for children. It was also a problem that arose in the historical development of mathematics. In mathematics education there is the expression "epistemological obstacle" to describe this kind of situation in which the historical process is replicated within the learning situation. The obstacle that arises in these two situations—learning and historical—are analogous precisely because they both involve a change in conceptual systems.

3.6 Zero

The examples of the integers and the fractions might seem to indicate that the new system, CS_2, must be much larger than the earlier one, CS_1, but, on the contrary, it is possible to enter into a new conceptual system with the addition of just one new number. The case in point is the addition of zero to the counting numbers or the fractions. The resulting system in either case is qualitatively different than the old one. It is easy to introduce zero as a computation in the system of counting numbers, say **3 – 3**. I have three apples and you take them all away and, as a result, there are no apples left. No problem here. This is very different from the difficult problem of conceiving of zero as a number on a par of with the other counting numbers. The reification of "nothing as something" was a feat that the Greeks never accomplished because they were offended by the very idea. The reluctance of the Greeks to embrace zero in this way is evidence that accepting zero is difficult in the way that we have come to expect when one shifts from one conceptual system to another.

The Greeks resisted the shift, not because they were stupid or deficient in some way, but perhaps because they understood that this would take them into a different conceptual world and they intuited that such a world would be at odds with many of the valued tenets of their culture. The Indian mathematicians did not have this reluctance. As I discussed in *How Mathematicians Think*, there were conceptual antecedents in the Indian culture and the Hindu religion that made the conceptualization of zero more acceptable.

3.7 Infinity

Infinity was another conceptual element that was problematic for the Greeks. They resolved the problem by a compromise that is interesting and instructive. They accepted the idea of an infinite process, as, for example, when you approximate a circle by inscribing and circumscribing it with a series of regular polygons with more and more sides. In the limit you can get as close as you wish to the area of the circle. You can use this process to approximate the numerical value of π and this "method of successive approximations" anticipates the fundamental idea of the calculus. Infinity as process was thus used in both computational and theoretical ways. On the other hand, "infinity as an object" was unacceptable to the Greeks and remained so to mathematicians and philosophers for two thousand years.

We can now understand what was going on in terms of our language of conceptual systems. "Infinity as process" is something that is acceptable within CS_1, the conceptual world of Greek mathematics, because you can think of it as a computation. It is a subtle computation, to be sure, but one which is essentially finite in nature. You may say that you can construct an infinite sequence of polygonal approximations to the circle but in practice you don't think of all of them at the same time. You work with one "sufficiently fine" approximation most of the time.

However "infinity as object" necessitates moving to another conceptual system and that is an enormous leap that neither the Greeks nor subsequent cultures were capable of making for a very long time. Again there are very good reasons for this. Progress is not merely a case of continuous improvement. You can't really develop the real numbers without thinking of an infinite collection (here of rational numbers) as one object. And if you do not do this then you can never develop the real number system, calculus, differential equations, and therefore much of physics and engineering. However making this leap means coming to live in a strange new world, which is unintuitive in many ways.[vi] The unintuitive aspects come to the fore in the work of Cantor but also in the earlier arguments about infinitesimals that marked the century after the introduction of calculus by Newton and Leibniz.

Actually using the word "unintuitive" for infinite objects gives away the game. It represents a marker that tells us that the conceptual system has changed and *the* defining aspect of such a shift is that there is an incompatibility between one's intuitions in the two systems. The development of the real number system from the rational numbers, as we shall see in the subsequent section, is a very good example about how vast a gulf may exist between successive conceptual systems.

3.8 The Real Numbers

I could have titled this section "the irrational numbers" where the word irrational would again highlight the obstacle to changing conceptual systems. The word "irrational" sends the same message as the earlier use of the word "negative."

However it is also interesting that the word "real" is used for the collection of rational and irrational numbers. Of course the "real" numbers have no claim to a deeper reality than the integers or the complex numbers. All number systems are "real" if you have mastered the appropriate conceptual system. The essence of any conceptual system is that it confers reality on the situation that it describes. The reality of a given situation is a function of the conceptual framework that the individual brings to it, which explains why teachers sometimes have great difficulty understanding their students' responses. What is "real" or "obvious" for the teacher is totally unbelievable for the student. Conversely, what is obvious for the student, from their standpoint, is often stupid or nonsensical for the teacher. The student's conceptual system is inevitably different than that of the teacher (or there would be nothing to learn) but different conceptual systems are incompatible with one another. A good teacher should be sensitive to this fact. The student is not wrong or stupid; she is living in a different mathematical world.

The irrational numbers come into mathematics as a result of Pythagoras' Theorem which gives you all of the square roots: $\sqrt{2}$, $\sqrt{3}$, $\sqrt{5}$, …. The problem with $\sqrt{2}$ is that it is incommensurable with the counting numbers. This means that it cannot be expressed as quotient of two integers. As a result of this, the Pythagoreans were forced to question one of their cherished beliefs, namely that all numbers were commensurable, that is, that the universe of numbers consisted only of fractions. This is the normal way that people feel about a conceptual system that they have mastered, namely that it defines what is real. The Greeks lived for a while in a world where all quantities were commensurable, that is, have a common measure, which is equivalent to saying that fractions are the only possible numbers. This point of view was extended to a more general belief, which many of us still share, that the entirety of the universe is "commensurable," that it is all tied together neatly by a rational framework, which can be described by mathematics. This is the belief in a rational and harmonious universe. The larger meaning of the fact that $\sqrt{2}$ is irrational (which may not have been lost on the Pythagoreans) is that, like it or not, discontinuities and incompatibilities are an irreducible part of the world. The world cannot be described by a single, rational conceptual system, scientific or otherwise, because you must deal with the problems that inevitably develop within any system and give rise to new systems.

From the modern point of view, that is, from the point of view of the conceptual system of the real numbers, square roots do not even come close to exhausting the set of all irrational numbers. I indicated earlier that an intermediate collection, the collection of "constructible" numbers, contains the square roots and forms a perfectly good number system. This is basically the system that defined "number" for the Greeks and it is an interesting conceptual system in its own right. Today we tend not to recognize this system because of our belief that number has its origins in counting. However this belief is erroneous—neither of the core systems that we discussed involves counting. There are other sources of number such as measuring, which are closer to the heart of Greek mathematics. People, like the mathematics educator David Tall, have argued, that measuring is also a sufficient basis for producing the number concept, albeit one that may produce different kinds of numbers. I would argue that the modern concept of number has its roots in the tension between counting and measuring.

As interesting as it is, the geometric system of constructible numbers has its limitations. This is seen in certain "unsolvable problems" that bedevilled the Greek mathematicians such as trisecting the angle or squaring the circle. These problems cannot be solved within the system of the constructible numbers. Their solution is dependent on an extension of the concept of number.

An even deeper problem concerns the number π. It is introduced as a ratio of the circumference of a circle to its diameter. The Greeks did indeed think of π as a number and not just a ratio and Archimedes succeeded in finding some good rational approximations. There were other numbers in the same category such as the "golden mean" that were clearly important but not constructible. What is the nature of these non-constructible, so-called transcendental numbers? There remains something mysterious and paradoxical about them to this day and this sense of mystery is precisely the feeling that arises when one's conceptual system is not completely assimilated.

We saw that the system of constructible numbers contains its own problematic elements, as do all conceptual systems. How does one develop a mathematical and conceptual system that contains the roots and also contains this other kind of non-constructible number? Before we get into a discussion of the modern answer to this problem—the real numbers—let me just note that there are many number systems that are situated in the intermediate domain between the fractions and the infinite decimals. For example, all numbers of the form $a+b\sqrt{2}$, where a and b are both fractions. (Or else replace 2 with any other non-square.)

The modern answer to the problematic elements of these various intermediate numbers system consists in making the huge leap to the conceptual system of the real numbers. Actually each number system has a problematic element, which eventually pushes us into a new and more complex system. $\sqrt{2}$ got us from the rationals to the constructible numbers. Numbers like π get us from the constructible numbers to the real numbers. The development of the real number system of infinite decimals is a relatively recent accomplishment (for mathematics) whose implications still remain unclear. Undoubtedly its development was a major intellectual triumph. The fractions, the roots and all the other irrationals including π and the golden mean can all be represented in decimal form. That is, all of the numerical systems that I have mentioned can be embedded in this new system and, as a result, their incompatibilities are reconciled. Like other mathematical breakthroughs this system formed the framework for a veritable explosion of new mathematics, especially the calculus and all of the scientific theories that depend on it. I shall return to a more detailed discussion of the strengths and weaknesses of the real number system in a later chapter.

3.9 Imaginary and Complex Numbers

The words “imaginary” and “complex” again demonstrate how difficult it is to make a major change in conceptual systems—a difficulty that we already encountered with negative numbers, fractions, zero, and irrational numbers. The word “imaginary” tells us that these numbers are unreal from the perspective of someone grounded in the real number system. I can remember having that feeling as a student and yet today I cannot recall what it was that made these numbers so seemingly illusory. It may have something to do with the fact that for most people "number" means a point on the number line. We don't carry around the full implications of the conceptual system of the real numbers so much as we carry around the simple geometric model of a line with each number occupying a precise location. From that point of view the number $\boldsymbol{i}$ whose square is $\boldsymbol{-1}$ is not a number since it can't be located on that line. I mention this because it will give most readers a concrete feeling of what it means to be grounded in CS_1 (the real numbers) and to be looking in the direction of what will hopefully become CS_2 (the complex numbers). We may not remember having trouble thinking of ***2/3*** as a number but most of us can remember having trouble thinking of $\boldsymbol{i}$ as a number.

As a computation over the real numbers, $\boldsymbol{i}$ is no great problem. We look at the equation $x^2 + 1 = 0$ and learn to solve it by means of the formula that gives you the roots of quadratics or, more directly, by treating this equation as an ordinary algebraic equation, arriving at $x = \pm\sqrt{-1}$ in the same way as we would solve $x^2 - 4 = 0$ to get $x = \pm\sqrt{4} = \pm 2$. When we write it out using the square root sign it even "looks" like a number. So in some ways it seems like a number—it is the solution of an equation and it has a notation involving square roots. On the other hand, it can't be a number because it has no place on the number line and we "know" that all squares are necessarily non-negative. Thus the situation is ambiguous and the resolution of this ambiguity involves the construction of an extended concept of number—two-dimensional numbers of the form $\boldsymbol{a + bi}$, where $\boldsymbol{a}$ and $\boldsymbol{b}$ are real numbers and $\boldsymbol{i^2 = -1}$.

How do we justify calling these new objects numbers? All complex numbers have the form $\boldsymbol{a + bi}$ and it turns out that you can do arithmetic with them, that is, you can add, subtract, multiply and divide them and keep the same form. For example, $\boldsymbol{1/(1 + i) = (1 - i)/2}$. It is possible to embed the real numbers in the larger system of complex numbers by identifying the real number ***3***, for example, with the complex number $\boldsymbol{3 + 0i}$. (Just like we embedded the counting numbers in the fractions by identifying ***3*** with ***3/1***.)

But these properties are not enough to justify calling the complex numbers a new conceptual system. The new system must also resolve some problem in the old. The problem that must be resolved involves algebraic equations like $x^2 + 1 = 0$ that we can state in the old system but not solve there. In the new system, not only can we solve this equation but also any (algebraic) equation with real coefficients. Even better, the system of complex numbers has the beautiful property that is called algebraic completeness, namely, that any (algebraic) equation that you can write down in the system of complex number also has (the right number of) complex number solutions. For example, the equation $x^3 + 1 = 0$ has the three solutions $x = -1$, $(1 \pm \sqrt{3}\, i)/2$. In this and many other ways, the complex numbers open up a new mathematical universe which not only has important properties and applications in itself but also reveals to you properties of mathematical objects that are obscure in the "real" world but become abundantly clear in the "complex" world. One important example involves the trigonometric and exponential functions—***sin(x)***, ***cos(x)***, and ***exp(x)***. As real functions, they seem to be unrelated but as complex functions they are seen to be so closely related, $e^{ix} = cos(x) = i\ sin(x)$, that one could really say that they are two different ways of looking at the same thing.[vii]

3.10 Abstract Mathematical Structures

Many of the number systems that have been discussed are examples of abstract structures. The rational, real, and complex number systems are examples of what is called a (mathematical) field. A field is an abstract structure in which one can do a generalized form of arithmetic. In other words, we have abstracted the essential computational properties of numbers, namely, that they can be added, subtracted, multiplied, and divided, and considered the most general situation in which these operations are possible. Actually the simplest such system consists of just two elements ***0*** and ***1*** which are added and multiplied in the normal way except that in this system ***1 + 1 = 0***. This is the system of binary, or ***mod 2***, arithmetic.

Field theory, like ring theory and group theory, is a sub-discipline of mathematics and there are mathematicians who spend their professional careers investigating the properties of these kind of algebraic structures.

The way these structures are normally introduced today consists of listing a sequence of axioms. This would include the property of closure whereby adding and multiplying elements of the system keeps you in the system, the existence of a unit, which would be the analog of the number ***1***, and another element, which would be the analog of ***0***. I will not write down the complete list of axioms because I am not interested in a field as an abstract structure so much as I am interested in a field as a conceptual system. There is a hierarchy of numerical conceptual number systems that I have described earlier and there is then a new conceptual system that abstracts their most important properties and is called a field.

To successfully operate within the theory of mathematical fields, one must rise up to this new conceptual level. Working at this level is thought of as working in an abstract setting as compared to working with an actual number system that most mathematicians would consider as being a concrete setting. What would be the distinction between abstract and concrete from the point of view of conceptual systems? The elements of a conceptual system that you have mastered feel "concrete;" when you are not quite there it may feel abstract. Actually no system of thought is really concrete; they are all abstract. It is more useful to think about going from one conceptual system to another.

Of course every proposition in field theory "contains" a whole series of propositions—one for every particular field. Each one of these (old) conceptual systems is one instance of the new system and some people operate in the abstract situation by simply translating every proposition into one or more "generic" examples. However there are others who develop a way of operating in the abstract system itself. They now have a "feel" for mathematical fields and have successfully established themselves within this new conceptual system. Nevertheless they maintain close touch with the "generic" examples that "represent" the abstract theory.

Every abstract mathematical system comes with such a conceptual dimension. You could say that a mathematical theory comes with these two perspectives: the objective "mathematical" perspective and the conceptual perspective. Only the first is normally considered actual mathematics but I prefer to think that every mathematical situation has these two perspectives. If one is interested in learning then it is essential to remember at all times that we are not merely dealing with an objective, abstract world but also with a conceptual world that has irreducible subjective aspects. The formal world only comes alive when it has such a cognitive dimension. It is only then that it deserves to be called real mathematics. When one mentions an abstract idea to a mathematician it evokes an entire constellation of concepts and ideas, what the mathematics educators Tall and Vinner[viii] have called a "concept image," the "total cognitive structure that is associated with the concept, which includes all the mental pictures and associated properties and processes."

3.11 The Discrete and the Continuous in Conceptual Systems for Number

The centrality of the notions of the continuous and the discrete, as well as the tension between them, can be found throughout mathematics and in all cultures. The work of developmental psychologists has indicated why this should be so. It has its origins in the two core systems for number that were discussed in Chapter 1. The "analog magnitude representation" system is continuous and approximate whereas the "parallel individuation of small sets" is discrete and exact. These two systems do not fit together seamlessly. There is an incompatibility there and one way to view the history of the number concept is that it consists of a series of resolutions of this ambiguous and problematic situation. We have two cognitive systems for processing number. In the absence of the other, each might be extended in such a way as to preserve its essential features, leading, perhaps, to two distinct number concepts but this is not the way that development happens. New systems are produced in order to resolve some of the discrepancies in the original systems or between two such systems.

As we saw in Chapter 1 the counting numbers, the first system that is learned, not only contains numbers that are larger than the ones processed in parallel individuation but also is capable of the precision that is lacking in the analog magnitude system.

The accomplishment of producing the system of counting numbers, as incredible as it is, is not the end of the line. The reason for this is that the tension between the discrete and the continuous has no definitive resolution. We saw how computations over the counting numbers lead to problems that cannot be resolved in that system and necessitate the creation of new number systems which are more complete and go part of the way towards reconciling the primal dichotomy contained in our core conceptual systems. Nevertheless, even if the systems we develop are increasingly sophisticated attempts to create one comprehensive number system, and to thereby pin down a meaning for number, the core systems remain in our brain and, as a result, the continuous and the discrete feel like different—even opposite—ideas. We live with the following situation: number is a singular idea—even if we can't quite say what it is—yet it is presented to us in two modes. It is this ambiguous situation that generated new mathematical systems in the past and will continue to do so in the future.

The reason that there is no definitive resolution to the tension between our core systems may well be because the core systems themselves evolved in such a way as to internalize distinct features of the natural world. This would happen in the same way that our organism has biological clocks that have internalized the twenty-four hour, circadian, rhythm that is so important to our survival on earth. Number, as almost all cultures have discovered and celebrated, captures some essential aspect of the world. You can call this aspect order, structure, pattern, or regularity. Number represents the intelligibility of the world without which the very survival of an intelligent generalist species like our own would be impossible. It is not surprising that we possess mechanisms for representing this aspect of the world. What is interesting is that we have more than one such system.

What is to be gained by having multiple representations for number? Perhaps the reason for them is to be found in nature, perhaps nature itself has more than one way of accessing those features of the world that are subsumed under the rubric of number? What would these features be? In my opinion, they are precisely the discrete and the continuous. Quantity, as Aristotle understood, comes in two varieties that he called magnitude, the continuous, and multitude, the discrete. We can think of this dichotomy as representing two elementary ways of introducing number into the real world—the activities of counting and measuring. But these activities do not merely go on in our heads—counting and measuring arise from features of the natural world. As I said earlier the world is ordered and a prime aspect of that order is quantity—the answer to the questions, "How much?" or "How many?"

Our brains have developed systems that replicated this ambiguous feature of the natural world. This parallel between features of the natural world and structures of our brain would go a long way towards explaining the central mystery of mathematics, namely, what has been called its "unreasonable effectiveness" in the natural sciences. Mathematics works because its origins lie in mental structures that evolved in such a way as to represent crucial features of the environment that human beings inhabit.

It is interesting that modern mathematics, pure and applied, still has not reconciled the differences between the discrete and the continuous. Even if number systems such as the real and complex numbers can claim to contain both continuous and discrete features, nevertheless it is still possible to divide the sub-disciplines of mathematics into those that are primarily discrete and those that are continuous. Calculus, differential equations, differential geometry and topology are continuous whereas number theory, algebra, and combinatorics are discrete. The pendulum swings back and forth as to which tendency is the dominant. In the days of Newton and classical mechanics the continuous seemed to capture the basic aspects of the universe whereas today, primarily due to the successes of quantum mechanics and the digital computer, there is a tendency to think of the discrete as more fundamental. However even subjects that are primarily discrete use continuous techniques and vice versa. The tension between the two remains a constant feature of mathematics. It enriches the subject and ensures that mathematics will continue to grow and develop.

3.12 The "Logic" of Deep Thinking

In the light of our expanded series of examples of mathematical and scientific conceptual systems, I move on now to a discussion of the manner in which the deep thinking that arises in the development and learning of such systems differs from the systematic, logical thinking that governs much of the work within a given system or paradigm.

Research on a conjecture or a problem in mathematics often proceeds in the following order:

Conjecture. You somehow come up with a conjecture on the basis of experience and intuition. You guess that the conjecture is generally true (or else guess that it is false) on the basis of partial evidence.

Idea. You get an idea of why it is true and how a proof might be put together (or an idea about how a counter-example might be constructed).

Logical Derivation. You attempt to write it up, that is, to derive the result from results that are generally accepted as true using a combination of formal and informal logic.

Difficulties. This initial attempt fails because of problems that were not thought of previously. At this stage, new examples may appear that either reinforce the original intuition (it's true) or changes the proposition to its opposite (it's false).

Change the Conjecture. At this stage one may change the conjecture or proposition by modifying the premises, conclusion, or both. Then the cycle of steps may be repeated.

This process continues until the original problem is solved or some sort of "interesting" proposition or example emerges. It always terminates at step 3, the proof or verified counter-example.

It may appear that deep thinking is not involved in this process at all but, in fact, it is essential. The process cannot begin without steps 1 and 2, that is, coming up with the conjecture and the idea for the proof. Deep thinking is often invisible in the completed proof, which appears to follow a linear, deductive sequence from beginning to end. Nevertheless deep thinking remains essential even when the result is proved and well established. The mathematician does not remember the proof so much as the idea behind the proof.

Even though the process of mathematical and scientific discovery consists of an alternation between insight and logical derivation, the latter seemingly has the last word. Thus many people have concluded that formal logic is the only way of using the mind that matters in mathematics whereas, in practice, what matters is getting the idea, the productive way of looking at the problem. Mathematics is not (merely) logical; its essence is deep thinking!

Deductive logic was invented by the Ancient Greeks and has become the foundation of all of mathematics and science. But logic is a language and a conceptual system in its own right with its own strengths and weaknesses. One of these weaknesses is that it tends to obscure the role of deep thinking in science. Deep thinking involves freeing the mind from its constraints—seeing beyond the current paradigm and reframing the situation. Formal logic, on the other hand, involves imposing a series of constraints. For example, the world of ideas and reality itself are massively non-linear and multi-dimensional but the world of logical discourse is linear. Deductive logic leads one to think about beginnings (hypotheses) and ends (conclusions). At its extreme this means finding the most elementary and minimal set of axioms on the one hand and a final deductive theory of everything, on the other. Deep thinking has no single location where it begins and it certainly does not have a definitive end. It is pure process—a journey with multiple stops along the way where one may sit and enjoy the view before taking up the creative task once more.

One of the most important theses of this book is that in deep thinking the mind is used in a manner that is fundamentally different from the way it is used in the everyday thought of adults but also from the way the mind is used in the analytic thought of scientific and mathematical theory and discourse. It is acknowledged today that there are many different kinds of intelligence[ix] but deep thinking is not just one variety of thinking. If thinking is inseparable from intelligence then deep thinking is intelligence of the most basic kind, the creative intelligence that is the subject of this book. It is built-in and this explains why it is even present in infants and in all situations of conceptual development and change.

On the other hand, formal logic, the normal rules of inference or predicate calculus, plays a key role in mathematics and the sciences. The role of logic is so central that many people would take the position that logic enunciates the objective rules of thought, in other words, that it is not possible to think correctly—or even to think at all—without the rules of logic. Such people would maintain that formal logic is the way things are—that logic is built into the world and our biology. If one believes this hypothesis it would follow that we should use the usual logical rules as *the* framework within which to discuss any aspect of reality and, in particular, deep thinking. I explained above, albeit in a cursory and incomplete way, why it is impossible to do so. To use the language of neuroscience: the brain comes with two quite different modes of intelligence (as we shall see in Chapter 4) and these modes are incommensurate with one another and autonomous—one mode cannot be reduced to the other. Logic and analytic thought constitute one of these modes; deep thinking is something quite different and involves both modes. As we shall see in the next chapter, one striking difference between those two modes is that the analytic mind is not capable of creativity. What is original only enters into the world by virtue of deep thinking.

Instead of trying to reduce deep thinking to logical thought, the problem is to explain how ordinary logical thinking arises out of deep thinking and then to describe the relationship between the two. Let us remember that formal logic came into existence out of an act of creativity—the ancient Greeks, people like Aristotle, invented it. But the rules of creativity—deep thinking—were evidently already functioning at that time. Deep thinking is not a cultural creation. Deep thinking is the way things are.

We take ordinary logic so much for granted that we have difficulty even imagining that there could exist a form of reasoning—much less any scientific or philosophical discourse—that was not structured in strictly logical terms. Thus the problem is this: How to have a discussion about formal logic without entering into the circularity of using the vehicle of formal logic to account for itself?

In what follows, I shall sketch out a different manner of thinking about the nature and role of logic. The first thing that must be done is to break the identification between generalized logic, understood as a way of reasoning or using the mind, and formal logic. Formal logic is systematic, linear, and incremental. It is a way of organizing analytic thought. But it is certainly not the only way in which the mind can be used; it should not be identified with the generalized process of reasoning. Thought can also be synthetic, non-linear, dynamic, and intuitive. The fact that such thought is beyond logic (trans-logical) does not imply that it is illogical.

I will entertain the risk of confusion by using the phrase "the logic of deep thinking" in order to stress that deep thinking is a way of using the mind that can be identified and learned (perhaps recaptured is a better way to put it). It appears in the context of conceptual development, learning, and paradigm change. It could also be called "developmental logic" or the "logic of creativity."

Our discussions in the last two chapters give us a hint of the natural distinction between formal logic and deep thinking. Recall the discussion of conceptual systems in Chapter 2. The most profound question that one can ask about conceptual systems is, "If a conceptual system defines what is real for the individual (or for society, for that matter) how does it change? How does one get out of one's current conceptual framework and go beyond CS_1?" We know that the creative way out is to develop a new way of seeing, a new conceptual system, CS_2. Though there is an obvious way to test whether or not a conceptual system has changed, the way that it happens will always remain a bit of a mystery. Even if we could (objectively) describe what happens the description would not be identical to the creative act. In other words the map is not the territory. However we certainly do know that it does happen and when it happens we remember the occasion as a moment of transformation when "everything changed." Conceptual change is the most significant and radical instance of deep thinking and one can see right away that its nature is completely different than that of deductive reasoning.

However deep thinking can also occur inside of a given conceptual system. Deep thinking is often the means by which some difficult problem is resolved—one manages to look at the problem and the situation that contains it in a new way. In the same way, the fundamental concepts that underlie a conceptual system are continually being understood in a new way or applied to novel situations. In each of these cases, deep thinking involves reframing the situation but this reframing can be local or global.

The processes that I have been talking about fall into two disjoint categories depending on whether or not they involve a fundamental reframing of the situation. If they do not then the process is usually governed by the rules of ordinary logic, by the principle of non-contradiction, for example. People tend to think that a proposition in mathematics is either right or wrong because they see the whole thing happening within a fixed conceptual system.

Deep thinking inevitably contains non-logical elements. When it occurs within a fixed conceptual space, such as proving a mathematical theorem, the result must later be reintegrated with the conceptual system of mathematics by means of a logical derivation. The well-established conceptual systems of science also obey the rules of ordinary logic and this leads many to make a fundamental error, namely, that the conceptual system comes into existence by only using ordinary logic, when, as I shall discuss in some detail in Chapter 5, the conceptual system arises through deep thinking in precisely those situations where the logical rules break down.

The kind of thinking that obeys the rules of ordinary logic at every step may be an essential step on the way to deep thinking. Such thinking is continuous, linear, incremental, and/or computational as in the arithmetic exercises we remember from our childhood. When the ideas and procedures are mastered, adding, subtracting, multiplying, and dividing can be accomplished in a mechanical way without the kind of tension that often accompanies situations that call for deep thinking.

A well-established paradigm or conceptual system is a comfortable place to be—when it works. You have a strong sense of orientation. This does not mean that you know the answer to every question that may arise but you have the feeling that the questions you face have answers and moreover you have a good feeling about how to go about getting the answers. You rarely have the feeling of being totally at sea which is the feeling that one has when one's conceptual system starts falling apart either as a result of experimental results that cannot be explained within the conceptual system or, in education, because a teacher has set the student a task that she knows cannot be successfully accomplished with the concepts and tools that the student has available at the time.

The difference here is, in a sense, the distinction between the continuous and the discrete—operating without reframing is continuous; reframing is discrete. Performing computations over a given conceptual system is a continuous activity as, for example, when we can divide ***5*** by ***3*** and say that the answer is ***1*** with a remainder of ***2*** without having to use the concept of a fraction.

Computations over CS_1 can be quite sophisticated and complicated but they do not change the nature of the conceptual system. There is an established conceptual system within which one performs the operations. We may develop a stronger intuition for the system because we have more experience of working with it in various ways. The framework may stretch but it does not break. The change in it is continuous. On the other hand, grasping a new conceptual system is radical and abrupt—CS_1 and CS_2 are incommensurable.

Continuous thought follows the rules of formal or informal logic. Thus we believe that a question cannot have two contradictory answers, that is, an answer is either right or wrong. That this is only so within a conceptual system and not between two different ones is illustrated by asking a child the question, "How many numbers are there between ***1*** and ***2***?" This question has different answers depending on the conceptual system within which the child is operating—the counting numbers or the fractions. Within a conceptual system one is bound by the need for logical consistency, which is the very thing that breaks down when the system proves inadequate to some of the demands that are put upon it.

CS_1 and CS_2 may both be areas of logical consistency but problems arise when we try to prematurely force a logical framework on the transition, that is, on learning, development, and creativity. Most descriptions of mathematics and science are descriptions from within a given conceptual system and therefore the whole system must be logically consistent. Even when we discuss the transition from one paradigm, CS_1, to another, CS_2, we often do so from the point of view of the more advanced CS_2 and so adopt a "view from within." As I pointed out earlier, we describe Euclidean geometry from the point of view of Formalism, a conceptual system that only arose as a result of the crisis associated with the advent of non-Euclidean geometries. Classical logic is the language and structural support of the view from within a conceptual system. We tend to think that it is neutral and, in doing so, we don't see that classical logic is itself a conceptual system—a way of viewing the world. When we adopt it we naturally inherit its consequences. These consequences include a partiality for objectivity, consistency, precision and intolerance for the problematic in the form of paradox, contradiction, and ambiguity. These latter elements, as we have seen and will see again in Chapter 5, have an essential role in deep thinking.

3.13 Conclusion

In these final words I shall go back to the list of properties of deep thinking that were initially listed in Chapter 1 and quickly review how they extend to the movement from one mathematical and scientific paradigm to another.

The first property I called "**naturalness**" which meant that deep thinking was the default setting of the mind. One of the reasons we have trouble accepting this is that we imagine that the default condition of the mind is one of quiescence, perhaps even a blank slate. However we know today that this is not true. The mind is naturally active and we can see evidence of this by measuring the activity of the brain. If the brain does not get a sufficient amount of external stimulation it will stimulate itself. When a baby (or an adult, for that matter) sleeps its brain does not turn itself off but engages in activities that are vital to its well-being. Thinking and intelligence are expressions of the natural vitality and dynamism of the mind. You don't have to turn them on; they arrive on the scene already turned on. Ironically it is the default condition of deep thinking that makes it so difficult to define. Deep thinking does not operate by way of objective precision. It is more basic than objectivity and precision which, as I mentioned above and shall go into in more detail in Chapter 4, are elements of a way of looking at the world that arises out of deep thinking.

Difficulty and **discontinuity** characterize paradigm change in the historical development of science and mathematics and are present in all of the examples I discussed. Thus deep thinking is often associated with controversy and crisis. In mathematics the problematic elements in paradigm change are obvious from the words that were historically associated with the development of the number concept: negative, irrational, imaginary, complex.

As was mentioned in the summary section of the earlier chapters, there may appear to be an inconsistency between the properties of "naturalness" and "difficulty." If it is natural how can it be difficult? The answer will turn out to be that there is something that blocks the natural functioning of deep thinking. We are captivated with the way we have come to see the world, in other words, with our operative conceptual system, and refuse to give it up. To move on our fixation with our present views must be broken down. This is always difficult and even painful.

Problems that can be stated in the language of one system often cannot be solved within that system because the solution depends upon the development of a new system. Mathematics is full of such examples such as when the properties of real-valued functions, for example, are most easily understood through considering the analogous complex-valued functions. With the victory of the new paradigm, CS_2, it may be hard to see what the fuss was all about. From the new point of view everything is integrated into one consistent and harmonious whole—we look at the world from our magnificent new paradigm and are excited about its possibilities and filled with energy and enthusiasm.

Uncertainty and Incompleteness. Uncertainty characterizes the situation of the scientist who has allowed herself to entertain a hypothesis that is inconsistent with or actually contradicts the operational paradigm. Psychologically this is a hard place to be in. The scientist is an adult and adulthood in most people is characterized by having internalized a certain view of the world. Scientists tend to have internalized the paradigms that were in existence during their formative years. (More of this in later chapters.)

All scientific theories are incomplete and approximate. They may function well at the centre of the domain that they describe but tend to break down at the boundaries. Paradigms are usually deep insights into some aspect of reality but they never capture reality definitively.

Reframing. Reframing is practically synonymous with paradigm change. It is deep thinking in action. You cannot make it happen nor can you concoct a formula, method, or procedure that will guarantee reframing. You have to be prepared to entertain two seemingly contradictory ideas in the mind and not flinch. Before it happens it is impossible; afterwards it is obvious.

Ambiguity. Paradigm change always involves ambiguity. As I discussed earlier, any significant concept has multiple representations and navigating between them requires a great deal of mental flexibility.

Reification. Reification refers to the manner in which hierarchies are established. Processes at one level become objects at a higher level. A good example involves programming and the way sub-routines are integrated into higher-level languages. Mathematics is replete with such examples that work both at the cognitive and also at the level of the formal mathematics where, for example, a real number can be represented as an infinite decimal.

Hierarchy and Integration. This is most obvious in mathematics but also evident in the sciences. The new system does not eradicate the old; the old remains around and needs to be integrated with the new. There is something very delicate going on in conceptual change that needs to be appreciated. On the one hand, the new and the old are incompatible with one another. On the other, the two systems can be integrated. This means that the person who has mastered the new way of thinking has the flexibility that goes with the ability to look at the situation in two or more different ways. It is important to note that not only are conceptual systems hierarchical but so are their objective projections into the natural world—atoms, molecules, cells, and so on.

Systematic Thinking. This is the kind of thinking that characterizes the period in which the implications of a given paradigm are being worked out. It is a period of great confidence and power when new problems are being resolved that formerly appeared to be impossible to resolve. Often there is the feeling that the paradigm is powerful enough to resolve *all* problems given enough computational power. Physics has found itself in that position repeatedly in its history. We are in that situation today with respect to the potential of the digital computer or in the biological sciences with respect to the potentialities of DNA research.

---

[i] This comment is usually attributed to the physicist Max Planck.

[ii] Eliade (1957), Chapter 1.

[iii] Low (1997), p. 101.

[iv] Byers (2007), pp. 153–157.

[v] Trudeau (1987), Chapter 8.

[vi] Discussed at some length in Byers (2007).

[vii] Byers (2007), pp. 225–231.

[viii] Tall, D. and Vinner, S. (1981).

[ix] Cf. Gardner (1985) or Miller (1984).

Chapter 4

The Mind and the Brain

4.1 Introduction

One of the features of creative intelligence and deep thinking is that they are basic and arise from a non-conscious level of the mind. It follows that it is impossible to describe them in a manner which is objective in the way one would define mathematical or scientific concepts. In the next two chapters we shall begin to see how and why this might be so. Wouldn't it be lovely if I could describe a simple method guaranteed to generate creative thinking or conceptual learning for that matter? Better still a method that I could describe in an algorithmic way. Unfortunately such a method would go against my repeated observations of the kind of difficulty that is inevitably present in deep thinking. The absence of a method that could be objectively formulated is a good part of that difficulty. (I recognize the objection that ChatGPT, for example, is capable of creativity but I'll postpone that discussion for a later chapter devoted to AI.)

Nevertheless it is possible to get some feeling for the manner in which the mind is used in the process of deep thinking. In this chapter I shall begin by returning to work on the thinking of babies as described by another developmental psychologist. Then I shall go on to describe recent work on the neuroscience of the brain to see in what way the structure and function of the brain can shed light on what is going on in deep thinking.

In the previous chapters I have discussed the development of number-based conceptual systems in children and the development of conceptual systems in mathematics and science. There were some striking parallels between these different developmental processes and my hypothesis is that these parallels exist because the process that is going on in both cases is a form of creativity and that any creative process often has a core of deep thinking.

Deep thinking is connected to consciousness but the connection is quite subtle and I shall address it in this chapter. Obviously deep thinking involves one's conscious mind but it is also evident that there are elements of it that are not conscious. This is a point that is stressed by Jacques Hadamard[i] in his treatise on creativity in mathematics and explains why insight, when it comes, often seems like it is discovered serendipitously and feels like it comes from sources that are outside of the (conscious) self.

We normally think that deep thinking, creativity and learning, are processes that go on exclusively in the human brain. On the other hand, the brain and the mind are themselves the product of evolution, which, when you come to think of it, is also an expression of creative intelligence. So the brain and consciousness are not necessarily a prerequisite to creativity but it is possible to maintain the converse proposition, namely, that the brain and consciousness both arise out of a process that follows the same general pattern as other creative processes. In other words, the evolution of brain and mind arise out of a kind of creative intelligence.

If deep thinking is such a basic aspect of things then we would expect to see it appear in the development of children at a very early age and, indeed, we have discovered that a kind of creative process is involved in children's learning of the number concept. We would also expect that our brain structure and function have evolved in a way that facilitates human beings ability to recognize, work with, and produce creative change. And, indeed deep thinking is there to be seen in our minds and our brains.

Deep thinking and creative intelligence are, I would argue, the essence of mind. It is not only the preserve of brilliant adults; it is not an add-on to human beings but the essential aspect of what it means to be fully alive.

4.2 Flashlight Consciousness and Lantern Consciousness

In this section we shall get a hint of how deep thinking is connected to consciousness—the way in which our minds interact with the world. This hint will come from additional recent research in developmental psychology on how babies and young children think. This will give us a sense of how deep thinking feels from the inside.

This new perspective will arise from some fascinating work in child development as described by Alison Gopnik, Professor of Psychology at Berkeley, in her book *The Philosophical Baby: What Children's Minds Tell Us about Truth, Love, and the Meaning of Life*. The relevance of Gopnik's book to deep thinking starts in her introduction where she describes the connection between human beings and change:

"The puzzling fact about human beings is that our capacity for change, both in our own lives and through history, is the most distinctive and unchanging thing about us. … What neuroscientists call plasticity—the ability to change in the light of experience is the key to human nature at every level from brains and minds to societies … Learning is a key part of the process, but the human capacity for change goes beyond just learning. Learning is about the way the world changes our mind, but our minds can also change the world. …"

Deep thinking is intimately connected with the ability to change and the ability to learn. Learning, "the way the world changes our mind," and creativity, the way "our minds … change the world," involve the same basic mechanism, namely, deep thinking and creative intelligence.

The best-known and most intriguing aspect of Gopnik's book is her discussion of consciousness. This is where we must look to get an idea about the way the mind functions in deep thinking as opposed to how it functions in normal adult consciousness.

She says, "All these differences between children and adults suggest that children's consciousness, the texture of their everyday experience of the world, must be very different from ours.[ii] … Is the way we see the world as adults the way we always have and always will see the world? Or could consciousness itself change?"[iii]

In fact it is clear today that consciousness evolves—both historically and developmentally. However it is important to remember that the earlier and more basic forms of consciousness do not disappear. They are always present but that does not mean that people are aware of them. Deep thinking involves accessing these more primitive ways of using the mind. Creative individuals, as we shall see in the next chapter, somehow have access to such modes of consciousness. In fact most people can be trained to be creative and this training will necessarily include accessing forms of consciousness that are characteristic of childhood.

Gopnik continues, "Babies are at least by some measures, *more* conscious than we are.[iv]… Babies and young children don't seem to inhibit distractions as well as we do—their attention is less focused. … So rather than determining what to look at in the world, babies seem to let the world determine what they look at. And rather than deciding where to focus attention and where to inhibit distractions, babies seem to be conscious of much more of the world at once. They aren't just picking up information about the specific objects that are useful to them—they are picking up information about all the objects around them, especially when that information is new."[v]

"This capacity for very general attention makes babies such terrific learners. … It's plausible that babies are actually aware of much more, much more intensely than we are. The attention spotlight in adults seems more like an attention lantern for babies. Instead of experiencing a single aspect of their world and shutting down everything else they seem to be vividly experiencing everything at once."

The distinction between "lantern consciousness" and "flashlight consciousness," gives direction to our search for the elements of deep thinking. The importance of this distinction is that it connects consciousness to learning and creativity. Learning and creativity involve using the mind in a certain way. But this way of using the mind does not just involve the normal sequential, verbal, or symbolic thinking that I described at the beginning of the first chapter. It is not logical and so many people would claim that it is not thinking at all. Yet it is certainly a mode of mental functioning, something that we all did quite naturally as babies but that we have forgotten as adults. It is therefore something very old and basic; yet simultaneously, it is, for most of us, something quite new. The baby is not picking and choosing what to attend to. It pays attention to what is present within its sensory field without censoring itself or making judgments.

Gopnik connects flashlight and lantern consciousness to brain processes when she continues, "there are two complementary kinds of brain processes that depend on experience in this way. Brains make more and more connections between different neurons, but they prune the less-used connections and retain only the most-efficient ones. Both these processes take place simultaneously throughout development. Both are shaped by external events. But the balance changes—earlier in life we make more connections, and as we grow older we begin to prune more connections. These processes may reflect complementary psychological processes, and even reflect the quality of our experience. Early in life we are sensitive to more possibilities, while later in life we just focus on the possibilities that are most likely to be important and relevant to us."[vi]

Making and pruning connections are examples of the two poles that are present in many different processes. These days technological change forces most people to continually master new ways of thinking and to simultaneously abandon old ways that have become inefficient or inappropriate. If we think that a conceptual system corresponds to a set of circuits in the brain then we can begin to see the neurological basis for deep thinking. A well-established paradigm or conceptual system would correspond to a set of brain circuits that are also well established and thus difficult to modify. This would be the physiological correlate to our observation that deep thinking and conceptual change are difficult. The individual tends to get stuck with those circuits that are deeply entrenched through repeated use. In order to make the leap to a new conceptual system the old circuits must lose their hegemony, that is, they must break down in some way. This breakdown will arise from a simultaneous set of conflicting demands.

Gopnik is not the first person to draw attention to the fact that consciousness can be experienced in these two different modes. She refers to the work of the great psychologist William James when "he says [that] the field of consciousness is like a narrowly focused beam with darkness all around it." For another, James says, "we may suppose the margin to be brighter, and to be filled with something like meteoric showers of images, which strike into it at random, displacing the focal ideas."

Gopnik goes on to point out that lantern consciousness is not reserved for infants: "Meditation and travel seem to end up causing what philosophers call the same phenomenology—the same type of subjective experience. … It's like the lantern consciousness of childhood as opposed to the spotlight consciousness of ordinary adult attention. You are vividly aware of everything without being focused on any one thing in particular. There is a kind of exaltation and a peculiar kind of happiness that goes with these experiences too."[vii]

4.3 Observing and Participating

However there is a further difference between lantern and flashlight consciousness that it is important to emphasize. This difference is pointed out by David Brooks in his book, *The Social Animal*,[viii] when he cautions that the lantern metaphor may seem to suggest that the infant is illuminating and observing the world and, in particular, that the observer is separate from what she sees. But, on the contrary, the infant isn't observing but immersing herself in the world. She is vividly participating in whatever captures her attention.

It is this distinction that I have in mind when I observe that consciousness might not be the right word for what is going on in "lantern consciousness." Normally consciousness is used in a transitive sense—one is conscious of something, that is, consciousness involves a dualism, a kind of observation. This would apply to flashlight consciousness but perhaps not to lantern consciousness. Low[ix] tries to get around this problem by using the word "awareness" and then splitting it into what he calls "awareness of" (flashlight consciousness) and "awareness as" (lantern consciousness).

Flashlight consciousness involves assuming the status of observer—standing outside of the situation. This is the only stance that you can take if you wish to analyze things. In other words, analysis is built upon flashlight, observational consciousness. Lantern consciousness, on the other hand, involves participation. Because participation does not require a clear differentiation between the self and the situation, it is a more intimate and more holistic form of awareness.

Participation and observation remain accessible to us as adults. When I sit in my study I can look at the details of the room—the books, the computer, the desk. I do this from the outside. Or else I can feel myself as part of the room, as part of something larger. I do this from the inside. So we observe from the outside and participate from inside. This is a vital distinction and it is well to recall that we all spent our early years in the lantern, participatory, mode and only later developed the observational stance that is so characteristic of adulthood. Synthesis, which I shall discuss next, has its roots in lantern consciousness and, for this reason, has an affinity to a more primitive way of thinking.

Normally intelligence is identified with the analytic and therefore with flashlight consciousness. However a baby is also intelligent—extremely intelligent in fact—yet her intelligence is clearly not the intelligence of flashlight consciousness. The point here is that deep thinking, the kind of thinking that is involved in creativity and conceptual learning, also involves more than flashlight consciousness and the analytic intelligence. Let's go on to explore the nature of these two ways of using the mind.

4.4 Analytic and Synthetic Thought

These days when students are asked to write a term paper on a given topic the first thing that they do is collect a mass of information using electronic databases such as Google or Wikipedia. There are really two steps here—collecting the relevant data, and then pruning and organizing the data. However, at some point, with a nudge from their teacher, students may realize that a term paper is more than a collection of facts; it must say something. It must be organized around an idea or thesis and must make an argument that supports the thesis. The thesis is essential to the paper, giving birth to its structure, providing meaning and coherence. This essential ingredient is often problematic for the student. "Where does the thesis come from?" she asks. "How do I convince the reader that my thesis makes sense?" A thesis serves the purpose of unifying the data.

It can usually be stated in one or two sentences yet it is the heart of the paper—without it you can't really even call it a paper but just a meandering flow of vaguely associated thoughts.

A computer is great for generating and organizing data but not so good at determining its meaning. A computer won't tell you what a book or a paper means nor can it reveal its central generating idea. The central organizing principle is the creative element in the term paper. Like all creative activity its origins are a little mysterious. It begins with a vague hint or feeling that there is something systematic going on within the situation that you are looking at even if at this stage you can't quite make that pattern explicit. Actually at this stage you often resist defining the idea too rigidly.

The idea often gets articulated in the writing process itself so at a given time one can later say to oneself, "Ah! So that's what I was trying to say." The thesis arrives suddenly when, instead of random facts, you are now looking at a unified body of information that makes sense. The thesis or idea that unifies the information involves synthetic thought. Synthesis, as the name implies, involves combining various elements into a whole. However, synthesis can be understood in a strong or a weak sense. In the weak sense the whole is nothing more than a vessel that contains the bits of information. However I shall use synthesis in a stronger sense: the whole is greater than the sum of its parts and its properties cannot be anticipated from the properties of the parts. In this way, words are not just collections of letters. Water is a synthesis of hydrogen and oxygen but its properties are "emergent," that is, they are not immediate from the properties of those two gases. Water is something entirely new! In just this way the Pythagoreans would not just see "three" as "one plus one plus one" but as something entirely new, something qualitatively different. Gestalt psychology emphasized the importance and elementary nature of "wholes" but we have trouble understanding what they were getting at because of the cultural dominance of analytic thinking. Synthetic thinking takes the wholes or unities as basic.

Let's go back to the term paper and point out that making a case that your thesis is correct uses a different way of thinking than the thinking that was used to produce the thesis in the first place. Making the case usually involves logical reasoning—what is required is a logical argument that will support the validity of the thesis. A logical argument starts with a set of assumptions, which are assumed to be true, and moves to a conclusion, the thesis, via a finite sequence of acceptable steps. In the best logical arguments the assumptions would be self-evidently true. In mathematics and in many other fields we usually call such self-evident truths axioms. This whole procedure is called analysis or the analytic method. It involves breaking things down to their most elementary constituents and then using logical argument to build up these atoms of thought into complex structures or conclusions. Scientific theories are often organized in this way. It is a powerful way of using the mind, which came into prominence with the mathematicians, scientists, and philosophers of Ancient Greece.

It is important to carefully distinguish between these two ways of thinking, in particular, to identify the appropriate role of the analytic and the synthetic in thought. Otherwise, one opens the door to fundamental misconceptions. One such misconception is the belief that a thesis arises as a result of a logical argument when this is not necessarily the case. The thesis usually exists before the argument has even begun since, without it, one would not know where to begin the logical sequence. In mathematics it is common to "think backwards", that is, begin with the conclusion and ask yourself what needs to be true in order for the conclusion to hold. What is the necessary condition?

Perhaps this confusion arises because the conclusion appears at the end of an argument. If one focuses on the argument and not on the underlying idea, we tend to think that the conclusion arises *because* of the argument. The logical sequence is not identical to the creative or the learning sequence. In research or learning, exploring the situation in various ways leads to the moment when one discerns a regularity, pattern, or structure in the elements of the situation, which is revealed in a unifying idea. The elements of the situation we are studying may well be inconsistent, ambiguous, or contradictory. The creative idea arises out of this incoherence, not necessarily out of the logical sequence or argumentation.

It is only after the thesis has come to conscious awareness that one begins the process of developing the formal argument. The formal argument is there for a number of reasons—communication, verification, clarification, and integration of the particular result with a larger body of theoretical material. However this should not obscure the fact that one often thinks in one direction and justifies in the opposite direction. Now thinking analytically may well reveal that the thesis is not quite right and needs to be modified but it will not tell you what modifications are necessary. The point here is that there are two ways of using the mind that come together to produce the term paper or any mathematical or scientific result. Synthetic thought is where the process begins and analytic thought arises at a subsequent stage. We shall see in the next section that these two ways of using the mind are mirrored in the structure and function of the brain. The brain appears to contain the potential for two distinct ways of experiencing the world, which correspond directly to the analytic and synthetic. The interesting thing is that these two ways of being in the world do not cohabitate peacefully. There is a tendency for the analytic to deny the existence of the synthetic.

Analytic thought begins with the details and builds these details into a larger description of the entire situation while synthetic thought begins with the big picture. When you encounter someone that you know, you don't see the details—arms and legs, nose and ears—you see George or Mary. When you hear a tune you recognize the tune as a whole and not the key or the individual notes. When you read you recognize whole words and can basically ignore the level of letters. This kind of "clumping" of information involves synthetic thought and is the basis of human cognition, how mathematicians think, and how computers are programmed. The details in each case disappear into the whole. We then are freed up so that we can just attend to the highest level because the hierarchy of more elementary levels is enfolded within it. However the details can be revealed, if and when we need them. Since our short term memories are only capable of holding a small number of independent facts at any one time the mechanism of clumping makes it possible for us to create the incredibly detailed structures that one finds in science and mathematics. This process of "clumping" whole sequences of thought together into units and giving them names is familiar from the routines and subroutines of computer programming but it is a quite general way of using the mind and explains how mathematicians and others are capable of holding in their minds arguments of extraordinary complexity.

What is the relation of synthetic and analytic thought to flashlight and lantern consciousness? Clearly analytic thought has an affinity to flashlight consciousness and synthetic thought to lantern awareness. Thinking involves both the analytic and the synthetic but there are deep and important questions about the relationship between them. Which is more fundamental? Are they independent of one another or can one of them be reduced to the other? Is there a conflict between them and, if so, how is the conflict resolved? Which one is more likely to produce creative ideas? Which one produces the truth?

In our culture analytic thought has the best press and is more easily understood. Society has great hopes for this kind of thinking. It is the hope that analytic thought can "capture" reality. An early exponent of this kind of approach to knowledge was Euclid with his invention of the deductive method, which, at a later date, was generalized from geometry to other fields. Ultimately it led to the dream of a "final" scientific theory that definitively captures all of reality in a one massive deductive system.

These dreams are nothing but the belief in the unlimited potential of analytic thought—that analytic reasoning is the way the mind works in general and the way in which the natural world functions. In this view of things synthetic thought can be reduced to the analytic. The unlimited potential of the rational, analytic process is a very powerful and very old project of Western civilization. For it to be realized we would have to answer the following simple questions about writing a term paper or dreaming up an interesting scientific experiment. Where does the idea for the thesis or hypothesis come from? How does the scientist know which one of the potentially infinite number of hypotheses to test? How can we say that one argument is "better" or more significant than anther? Why is a particular result "deep" or "elegant" or even "beautiful"? Some people think that these questions have no answers because they are "subjective" but, on the contrary, questions of significance and beauty are of immense importance, especially, in science. When we say they have no answers we really mean that there are no answers from a purely logical and analytic point of view. The "answers" exist but come from using the mind differently and opening oneself up to synthetic as well as analytic thought.

It's a bit like the question that students invariably ask when presented with some complex and ingenious mathematical argument. "How did she (the mathematician) come up with the idea for the proof?" "How can I hope to come up with such an idea or *any* idea, for that matter?" This is a variant of another observation that students make when shown such an argument. "I follow it (the reasoning) but I don't understand it." "Following it" is an exercise in analytic thinking. (I will go propose a another way of thinking about rational thought in the chapter on the nature of mathematics which will incorporate elements of analytic thought, intuition, and insight or deep thinking.)

With enough work students can be trained to master analytic thought but I must admit it is getting increasingly difficult to get some of them to think in this way which feels alien to the social media generation. However, "understanding" is more subtle than analytic thought. It means getting the idea that stands behind the argument, the idea that produced the argument in the first place. This is the creative core of learning—what real education is all about. "Understanding" is a creative process that is discontinuous—you either understand it or you don't—and there is no algorithmic way of attaining understanding. Understanding and coming up with an idea are both creative processes that involve both the synthetic and the analytic.

Analytic thought is familiar to most of us—the more educated you are the more familiar it is. Synthetic thought is more mysterious precisely because it is not a function of analytic thought or even of normal adult consciousness. Its origins are deeper and more primitive than that. People with very low IQs, so-called idiot savants, are capable of extraordinary feats of memory, mathematics, art, and music because they use different mental capabilities than the ones we normally associate with intelligence. They "know" without conscious calculation that a five-digit number is or is not a prime or that ***111*** decomposes into three copies of ***37*** even if they are unable to multiply or even conceive of multiplication.[x] This is a different kind of "knowing" than the one we associate with reasoning and analytic thought. It is a synthetic cognition rather than analytic. Reading Gopnik reminds us that synthetic thought has strong similarities to the kind of awareness we had as infants before we learned the analytic and focused consciousness within which most of us spend our waking adult lives.[xi]

4.5 The Asymmetrical Brain

The significance of the pairs of mental dichotomies that I have been discussing—the analytic versus the synthetic, observing as against participating, or flashlight consciousness in comparison to lantern awareness—are highlighted by recent work in the field of neuropsychology. Different ways of using the mind naturally correlate with differences in brain structure and function. This does not necessarily mean that the structure of the brain *causes* these different modes of mind, which is the usual way of looking at the relationship between brain and mind. It merely means that there exists a systematic correlation between the domains of mind and cognition, on the one hand, and brain structure and function, on the other. "Isomorphism" is the mathematical word for this kind of correlation. Of course in mathematics the notion of isomorphism carries the precise meaning of a one-to-one mapping from one mathematical structure to another, which preserves some essential element of the structure such as when a group isomorphism preserves the multiplicative structure, $f(a \times b) = f(a) \times f(b)$. If I say, "mind is isomorphic to brain," I am not thinking of any explicit mapping between the two domains so much as I am pointing to certain deep analogies between mental processes and brain structure and function.

This book is in a fundamental way a meditation on isomorphic processes. I suggest, for example, that development is isomorphic to learning and creating. Or, as is highlighted in the notion of epistemological obstacles, that the learning of mathematics recapitulates significant events from the historical development of the subject. The way to think about such things is that we have two different perspectives on the same situation. For example, there is a common process that underlies both child development and the more systematic learning that happens at school. We might also think about an isomorphism as a description of the same situation in two different languages or from different points of view.

Calculus gives a mathematical example of such a dual perspective where the differential and integral calculus are the two perspectives. The isomorphism in this case is given by the Fundamental Theorem of Calculus, which enables one to translate statements from one language, the language of derivatives and differential calculus, to the other, the language of integral calculus. This allows you to solve difficult problems in one domain by translating them into the other where they may be easier to resolve. This is the essential value of constructive ambiguity. It allows you to "reframe" the situation and this gives you a new power and flexibility.

In this section I am concerned with two different and fundamental ambiguities—brain and mind, on the one hand, and left and right hemisphere, on the other. We have one unified brain that has a deep hemispheric division. This gives human beings two ways to think and to be in the world. What is the evolutionary advantage of having a bifurcated brain? It is not merely a matter of building in redundancy for then the two hemispheres would be symmetric in structure and function, which is not the case. There must be an evolutionary advantage to having this mental duality. What could it be? Perhaps nature makes the evolution of an ambiguous brain structure necessary or advantageous.

I have discussed various kinds of ambiguous structures and processes in this book. Each ambiguity consists of one unified situation that can be interpreted in two, conflicting ways. We saw how the need to reconcile problematic situations was a motor that generates development and learning and we shall see that it is a key element in creativity in mathematics and science. Perhaps the existence of dual processing units is the biological motor that makes human life into the dynamic, creative, and developmental process that it is.

The 1960's saw a lot of enthusiasm for a simplistic interpretation of the left/right brain split. Language was on the left, geometry on the right and various activities were characterized as left or right brained. However it now appears that language, mathematics, science, and creativity in general involve both hemispheres. Accordingly the research today is much more subtle. Because it has such an important bearing on my thesis, I shall spend a little time discussing what modern research has to say. My primary source is a fascinating account: *The Master and his Emissary*[xii] by Iain McGilchrist, a psychiatrist, philosopher, and Fellow of All Souls College at Oxford. He is part of a growing number of people who have revisited some of the traditional questions in philosophy from the point of view of the empirical results of psychology and brain science.

McGilchrist's thesis is, "that for us as human beings there are two fundamentally opposed realities, two different modes of experience; that each is of ultimate importance in bringing about the recognizably human world; and that their difference is rooted in the bi-hemispheric structure of the brain." "Why," he asks, "is the brain so clearly and profoundly divided? Why, for that matter, are the two cerebral hemispheres asymmetrical?" In his search for an answer he points to the distinction between synthetic and analytic in terms that are very reminiscent of my earlier discussion: "One of the more durable generalizations about the hemispheres has been the finding that the left hemisphere tends to deal more with pieces of information in isolation, and the right hemisphere with the entity as a whole, the so-called Gestalt … there are two fundamentally different 'versions' delivered to us by the two hemispheres, both of which can have a ring of authenticity about them, and both of which are hugely valuable; but that they stand in opposition to one another, and need to be kept apart from one another—hence the bi-hemispheric structure of the brain."

It is this "opposition" that is the key to understanding hemispheric differentiation. We have seen that "opposition" is the reason that deep thinking has an intrinsic difficulty, which can arise in the form of ambiguity, contradiction, or merely as dissonance or conflict. I insist that opposition, as opposed to mere difference, is the crucial factor precisely because it is opposition, and the need to overcome it, that underlies the dynamism of our mental life and of development in general. Such opposition does not arise as a result of a mistake but is the essence of life. "The brain is, in one sense, a system of opponent processors. In other words, it contains mutually opposed elements whose contrary influence makes possible finely calibrated responses to complex situations."[xiii]

"There are two different ways of using the mind that arise out of the disparate functioning of the two cerebral hemispheres. We have two relatively complete cognitive apparatuses. It is not that the hemispheres are different modalities from which to look at a pre-existing world but that each modality is a distinct way of being in the world. In a way each hemisphere brings a different world of human experience into existence."

It is the conflict between these two modalities that is crucial to McGilchrist. His book is meant to bring this conflict out into the open and explore its implications for western culture, past and present. It was this stress on the brain's two conflicting modes that aroused my interest in his book because it seemed to provide empirical support for the point of view that I have been taking about development and creativity. I claimed that every conceptual system is a window on reality and that two such systems inevitably conflict because reality is always identified with a particular conceptual system. If the two cognitive systems supported by our two hemispheres produce two different worlds then this would lead to great discord, in the first instance. Integration and harmony would only come later, if at all. We would have to work hard in one way or another to produce a higher-level synthesis. It is this kind of difficult activity that I have been associating with development, learning, and creativity.

This view is very different from the normal one in which we take for granted that the integration of the activities of our cerebral hemispheres is itself built-in. We tend to be put off by the idea that human beings carry around a conflict within themselves[xiv] and that much of life and culture, including science and the arts, are attempts to resolve this conflict. Fortunately or unfortunately, irreducible conflict is the way things are.

Various types of resolutions are possible.[xv] My book is about constructive kinds of resolutions that occur through creativity, development, and learning. Other, more destructive, resolutions are also possible and arise from suppressing one aspect of mind in favor of the other. When projected out into the external world it leads to the endless struggles of the so-called "good" for a definitive victory over the "bad" or the "other." This battle can also be framed as the "light" against the "dark", or the "pure" against the "impure". Framing a situation in this way makes the conflict appear to be irreconcilable and so often degenerates into violence.

This dynamic of suppression can also be found in mathematics and science where, hopefully, conflict does not lead to violence (although let's not forget episodes like Galileo and the Church). What one observes in science and mathematics is a powerful tendency to impose a "left-hemisphere" view as real (and I shall describe below what this means and implies). This would include seeing the brain as a kind of computer or viewing mathematics in strictly formal, deductive terms.

Situations where we have two conflicting views have come up in other places in this book. For example, there are, according to Carey, two core conceptual systems for number. Why two and not just one unique system? Possibly because the two hemispheric systems look at "number" differently, and indeed, the characteristics of number that I detailed in Chapter 1 fit nicely with the characteristic modes of cognition of the two hemispheres as McGilchrist describes them. This is another bit of evidence that the differences between the core systems arise out of the divided brain and that subsequent development is pushed along by the need to reconcile these differences.

As we shall see in the next chapter, various researchers in the field of creativity hold that the vital step in the creative process involves holding two contradictory ideas in the mind simultaneously. Thus creativity depends on duality and conflict. In this section we are discussing how the brain also contains duality and conflict that these arise from its bifurcated structure and function. This implies to me that the brain is set up to be an organ of creative development and learning. The brain is not set up to be a camera and take snapshots of reality. It is designed to facilitate creative development and learning and to do that you require different points of view that conflict with one another.

This is the kind of instrument with which nature in her wisdom has provided us. However I want to make my perspective perfectly clear and repeat that the structure of the brain does not necessarily "cause" analytic and synthetic thought or any of the other dichotomies that I have been discussing. I maintain that the same underlying situation that leads to different ways of using the mind is also reflected in the structure and functional specialization of the brain. It is not that "mind causes brain" or that "brain causes mind" but that both brain and mind arise out of the process of creative intelligence that I have been documenting in this book.

If it is indeed true that our brains are divided in this way, then it is not entirely good news. Both parts have a claim to be "me" but, of course, there can only be one "me." One might argue that the two hemispheres communicate through the corpus callosum and so cooperate with one another and this would be true to a certain extent. But it is also true that a prime function of the corpus callosum is to *inhibit* communication! It is this inhibition that makes it possible for each of the hemispheres to produce a stable and consistent view of the world. There is a unity underlying these two views but it is often implicit not explicit, that is, we are not necessarily conscious of it. At the conscious level we often identify with one view—in "modern" people this would be the analytic view that is characteristic of the left hemisphere. Often the other view is repressed or denied.

This inner division is the source of humanity's greatest triumphs but it is also why human life is so difficult. Our very nature calls out for a resolution to this duality and we are all driven in this direction. Creativity is a way of answering this call. Learning, creativity, and development are not luxuries but necessities. Without a creative resolution one is left with the abortive pseudo-resolution of repression and violence. But make no mistake about it, one way or another this deep schism will generate some sort of resolution. The choice is really only between whether the nature of that resolution will be constructive or destructive.

4.6 The Worldview of the Hemispheres

When a flock of Canada geese is feeding, one or two of the adult males stand guard in an upright, attentive posture which leaves the rest of the flock free to focus on feeding. This is a good example of the two types of attention that most animals including humans can bring to bear. It is a rudimentary form of what I earlier called focused and unfocused awareness. Each of these types of awareness is a characteristic of one of the two hemispheres. Both types of awareness are necessary and important. On the one hand, we must be able to focus on the task at hand and ignore anything that is not immediately relevant. On the other, we must be on the alert for danger and opportunity; we must be open to what is new and unexpected.

McGilchrist says, "In general terms, then, the left hemisphere yields narrow, focused attention, … The right hemisphere yields a broad, vigilant attention, …" and he continues, "In humans …each hemisphere attends to the world in a different way—and the ways are consistent. The right hemisphere underwrites breadth and flexibility of attention, where the left hemisphere brings to bear focused attention. This has the related consequence that the right hemisphere sees things whole, and in their context, where the left hemisphere sees things abstracted from context, and broken into parts, from which it then reconstructs a 'whole': something very different." This description should ring a bell because it is reminiscent of my earlier discussion of the characteristics of mind: analytic and synthetic thought or flashlight and lantern consciousness.

He summarizes, "Hence the brain has to attend to the world in two completely different ways, and in so doing bring two different worlds into being. In the one, we experience the live, complex, embodied, world of individual, always unique beings, forever in flux, a net of interdependencies, forming and reforming wholes, a world with which we are deeply connected. In the other we 'experience' our experience in a special way: a 're-presented' version of it, containing now static, separable, bounded, but essentially fragmented entities, grouped into classes, on which predictions can be based. This kind of attention isolates, fixes and makes each thing explicit by bringing it under the spotlight of attention. In doing so it renders things inert, mechanical, lifeless. But it also enables us for the first time to know, and consequently to learn and to make things. This gives us power."

This description of the predilections of the right and left hemispheres is especially important to any discussion of mathematics and science. For example, the formalist description of mathematics has the effect of reducing mathematics to an exclusively left hemisphere activity. Our job, as philosophers of mathematics, as mathematicians, and as educators is to comprehend how the two hemispheres collaborate to produce mathematics and learning. In doing this we must be aware that there are these two possible ways of approaching mathematics and that there is a tension between them which results in one mode, the analytic, denying the importance and even the existence of the other, synthetic mode.

This explains a great deal about the history of mathematics. It has been clear for a long time now that formalism does not describe mathematics as the working mathematician experiences it. There is a huge gap between what a mathematician does and what he says he does. What he does is creative work in which intuition and rigour, synthesis and analysis all have an important place. What he says he does is prove theorems,[xvi] which is nothing but a reduction of mathematics to a left hemisphere point of view. Now mathematicians can get along quite well in their research efforts with saying one thing and doing another but philosophers cannot and neither can teachers and students.[xvii]

4.7 Learning and Creativity Start and End in the Right Hemisphere

The right hemisphere is where learning and creativity begin: "…in almost every case what is new must first be present in the right hemisphere, before it can come into focus for the left.… Not just new experience, but the learning of new information or new skills also engages right-hemisphere attention more than left, even if the information is verbal in nature. However, once the skills have become familiar through practice, they shift to being the concern of the left hemisphere, …. If it is the right hemisphere that is vigilant for whatever it is that exists 'out there,' it alone can bring us something other than what we already know. The left hemisphere deals with what it knows, and therefore prioritizes the expected."

Learning and creativity originate in the right hemisphere because only the right hemisphere can recognize novelty and gestalts (or wholes). It only moves to the left hemisphere subsequent to being recognized by the right. This explains a great deal about conceptual development and creativity that is otherwise opaque and mysterious. A good example comes from my discussion of "number." We seem to "know" number before we can pin down that knowledge by means of a formal definition but that initial knowledge is often dismissed as informal or intuitive, as merely involving an imprecise "feeling." The reason for rejecting the more informal concept is that it lacks precision and therefore cannot be used in any subsequent analytic process. However this is just another way of saying that the concept starts on the right side and only later moves to the left. Precision and logic are characteristic ways in which the left hemisphere operates.

In a textbook or research monograph the concept is identical to its denoted meaning; the broader concept image or range of connotations is ignored. The reason for this is that the process of writing down mathematics and communicating it to others only seems to be possible after a precise conceptual framework has been laid down. How can you prove theorems about continuous functions if you don't have a precise definition? And yet even the absence of the famous ε–δ definition does not mean that we have no "feeling" for what it means for a process or a geometric curve to be continuous. We begin with a very definite intuition for continuity, which comes from the right hemisphere but may well be embedded in some core cognitive system just as we have core systems for number and geometry.

What distinguishes a "good" definition from one that is essentially useless? The famous physicist Eugene Wigner[xviii] claimed that the reason that mathematics was so "unreasonably effective" in the natural sciences was precisely the subtlety and ingenuity of its definitions. A definition is only valuable to the extent that it captures significant aspects of the right-hemispheric notion that is connected to our experience of the natural world. A definition is not arbitrary, as you might sometimes conclude from a formalist reading of mathematics.

Many of the building blocks of science—concepts like number, randomness, time, space, matter, and energy—cannot be captured definitively by a formal definition. This does not mean that they cannot be defined. Merely that no ultimate or final definition is possible. Such incompleteness makes sense from the point of view of hemispheric differentiation. The left hemisphere cannot "capture" the view from the right and cannot create its own original view—it can only "re-create" a view that originates in the right hemisphere. To do this it must begin by translating the right hemisphere's version of the concept into terms that it can deal with and this includes making processes into precisely defined objects or connoted meanings into denoted meanings. There are often many ways of translating the right hemispheric notion into the left's domain and this would mean, in particular, that different definitions of number are possible (as we have seen). Each possible definition captures one or more aspect or aspects of number and gives rise to a mathematical theory and corresponding conceptual system.

Number is known first by the right hemisphere and then transferred to the left. The right hemisphere's version is inevitably imprecise and "unknowable" by the left hemisphere. The same considerations tell us something about the relationship between formal and informal mathematics. To the extent that we identify the informal and intuitive with the right hemisphere and the formal with the left, it is necessary to have an informal feeling for a subject before we can work with it in a meaningful way. The "right way" must precede the "left way" or else we are left with a series of mechanical derivations that have no significance. The priority of the right hemisphere explains why we sometimes have an intuitive sense of the correct direction to proceed before we can justify this sense though a logical argument. "Give me the basic idea," most mathematicians would say, "and I'll fill in the (logical) details."

Most of the thinking we do in mathematics (and elsewhere) involves performing computations within a well-known paradigm, CS_1, a process which, roughly speaking, involves the left hemisphere. The leap to a new idea or new conceptual system, CS_2, inevitably utilizes the right hemisphere's ability to recognize novelty and so the creative step involves returning the situation to the right hemisphere. In general we can only understand creativity and learning by invoking the alternation of our two hemispheric functions. If we attempt to reduce learning or creativity to an exclusively left-hemispheric activity we will inevitably miss its essence. If our teaching is only directed to the mastery of left-hemispheric skills then things like understanding will be left to develop in a haphazard way at best and at worst will be seriously impeded.

McGilchrist summarizes the total procedure as follows: "What is offered by the right hemisphere to the left hemisphere is offered back again and taken up into a synthesis involving both hemispheres. This must be true of the processes of creativity, of the understanding of works of art, of the development of the religious sense. In each there is a progress from an intuitive apprehension of whatever it may be, via a more formal process of enrichment through conscious, detailed analytic understanding, to a new, enhanced intuitive understanding of this whole, now transformed by the process that it has undergone."[xix]

4.8 The Left Hemisphere's Tendency to Rationalize

I have repeatedly noted that deep thinking—creativity and conceptual development—is hard. We are now in a position to see why this should be so for the difficulty is rooted in the left-hemisphere's tendency to resist change. In other words, we become attached to the conceptual system or point of view with which we are familiar. Breaking free from this state of inertia necessitates an extraordinary effort. The nature of the kind of effort that is required is what makes creativity and conceptual learning difficult. It is not enough to work hard, to huff and puff, or spend enormous amounts of time. I can remember a student who complained bitterly that he had spent many hours working on a mathematical homework problem without any success and felt that he should be given credit for the effort he had put in. I asked him to tell me what exactly he had been doing for all of those hours. It turned out that he worked up to a certain point, got stuck, returned to the beginning and repeated this sequence over and over again. Repeating some procedure repeatedly will never lead you to something new but the student had no idea of how to get out of the rut in which he found himself.

There have been many experiments with split-brain patients that point to the left hemisphere's propensity to rationalize, that is, to make up explanations that have no basis in reality. We all have had the experience of making up an excuse for something out of thin air and have seen this kind of thing in our children when they are accused of some minor transgression—"the dog ate my homework" kind of excuse. Politicians use this capacity of the mind when they hire an army of "spin-doctors" who are prepared to invent an instant explanation for every mistake or inconsistency in their candidate's positions. The tendency to rationalize in this way is so prevalent because it is a basic mode of functioning for our left hemisphere. Evidently the tendency to rationalize is not restricted to human being but extends to other primates as well.[xx]

McGilchrist describes hemispheric differences in facing a problem in the following way: "It is similar with problem solving. Here the right hemisphere presents an array of possible solutions, which remain live while alternatives are explored. The left hemisphere, by contrast, takes the single solution that seems best to fit what it already knows and latches onto it. V. S. Ramachandran's studies of anosognosia (in which a person who suffers from a disability seems unaware of the existence of his or her disability) reveal a tendency for the left hemisphere to deny discrepancies that do not fit its already generated schema of things. The right hemisphere, by contrast, is actively watching for discrepancies, more like a devil's advocate. These approaches are both needed, but pull in opposite directions. ... The left hemisphere operates focally, suppressing meanings that are not currently relevant. By contrast, the right hemisphere processes information in a non-focal manner with widespread activation of related meanings."

Most of our work is inevitably based on some well-understood paradigm or conceptual system. Because it is familiar it resides in the left hemisphere. This facilitates the kinds of computations that I have spoken about. These computations occur within a fixed conceptual universe and continuously and incrementally extend the boundaries of that universe. What does one do when one encounters anomalies within the operational paradigm? Well the tendency of the left hemisphere is to reject discrepancies. This makes Carey's question about how we ever get from CS_1 to CS_2 even more important and mysterious. Given what I have said about the left hemisphere how *do* we manage to learn or create anything new?

New conceptual learning can only occur after the individual has escaped from their operative mindset. This explains in large part the nature of the difficulty that is associated with such activities. How does one go about letting go of one's mindset when the attempt to do so precipitates a situation that is almost paradoxical? The habitual mindset is lodged in the left hemisphere, which does not want to let go. Any attempt to force it to let go, even to focus on letting go, just reinforces the control of the left hemisphere. How do we escape from the tyranny of the left hemisphere and let go of flashlight consciousness? (In Chapter 5 I shall discuss the paradox of "letting go" in greater detail.)

The answer to this question is that the left hemisphere's tendency to be fixated on the known must be broken down. The way to do this is to learn to work in the opposite direction from that of repressing the problematic elements of the situation. On the contrary, it is the ambiguities and the contradictions with their tensions, inconsistencies, and seemingly irreconcilable conflicts that provide an opportunity to break down the fixation with the known. This is the purpose of "holding two contradictory ideas in mind" which we will discuss in the next chapter as the "doorway to creativity." It is only dissonance that can break up our current mental framework and open the gates to something new. Logic and other kinds of continuous computations will never change our mental set. This is my problem with those models of the mind/brain that are based on the computer. Creativity is a process that entails transcending one's current mental model. It does not happen by focusing more intensely or by extending the range of phenomena that are under the control of left hemispheric processes.

One does this by authentically confronting the problematic elements of the situation, which may be an ambiguity, paradox, or contradiction or merely the feeling that something is wrong and does not make sense. That is, one absorbs oneself in the situation in all of its complexity setting aside as much as possible those inner demons that tell you *a priori* what is possible or reasonable.

Having gone as far as one can at that time one has to have the courage to walk away and put the problem down for a while. This may mean going for a walk or getting a coffee, or just going to bed. You give yourself permission to put the burden down because you know that other parts of your mind will not stop functioning. One's attention goes from being focused to the sort of unfocused awareness that was called lantern consciousness. In that pregnant state any random event that is picked up by the right hemisphere can precipitate the creative association. It is precisely the right hemisphere's ability to put together different kinds of combinations, to juxtapose diverse experiences that ostensibly do not belong together that ultimately brings about the insight that appears to the rational mind to arrive serendipitously.

McGilchrist says, "It may be that cessation of the effort to 'produce something'—relaxation, in other words—favors creativity because it permits broadening of attention, ... This explains the 'tip of the tongue' phenomenon: the harder we try, the more we recruit narrow left-hemisphere attention, and the less we can remember the word. Once we stop trying, the word comes to us unbidden."

4.9 Implications for Deep Thinking

This discussion has immediate relevance for my discussion of deep thinking and of thinking in general. There are different forms of thinking and corresponding kinds of learning (as we shall see in a later chapter). Now we can see that many of their distinguishing characteristics follow from the dominance of one of the cerebral hemispheres. Creativity and conceptual learning do not follow from merely rearranging predetermined elements. Thus a computer program or a mathematical proof may be a good model for the functioning of the left-hemisphere but not for the integrated mental functioning that is required for deep thinking.

The left/right dichotomy evidently also has implications for what we consider to be the real nature of mathematics (and science). The view of mathematics from the left hemisphere would naturally emphasize its axiomatic, logical, and formal aspects. What is mathematics when viewed from the right hemisphere? It might well be the approach to mathematics that I have talking about in my books. It is a view where the left hemisphere is an indispensable contributor but where the right hemisphere has the first and the last word.

McGilchrist says, "Certainly there is plenty of evidence that the right hemisphere is important for creativity, which given its ability to make more and wider-ranging connections between things, and to think more flexibly, is hardly surprising. But this is only part of the story. Both hemispheres are importantly involved. Creativity depends on the union of things that are also maintained separately—the precise function of the corpus callosum, both to separate and connect; and interestingly division of the corpus callosum does impair creativity."

4.10 Conclusion

The creative intelligence that we access through deep thinking is neither analytic nor synthetic, neither right-hemispheric nor left. It arises from a complex interaction between the two. Real science is "deep science"; real mathematics is "deep mathematics." It involves both hemispheres, both analytic and synthetic thought, but the life of the subject—creating and learning—resides in the right hemisphere and in synthetic thought. It is not the current theoretical picture that is "true" so much as it is the process of doing and understanding. The current model is always out of date; the new theory is continually being born. (I shall devote a later chapter to the implications of losing at mathematics from a "process" point of view.)

Unfortunately we take the analytic view to be the only reality and so we distort what we are attempting to describe. We have a deep need for a viewpoint that does not change; we wish to pin down the truth. If we wished to develop a view of mathematics and science that unified the activity—the "doing" of science—with the theoretical content then that view would necessarily involve the perspectives of both hemispheres.

---

[i] Hadamard (1954), Chapters 2 and 3.

[ii] Gopnik (2009), pp. 13–14.

[iii] *Ibid.*, p. 16.

[iv] *Ibid.*, p. 110.

[v] *Ibid.*, pp. 118–119.

[vi] *Ibid.*, pp. 121–122.

[vii] *Ibid.*, p. 129.

[viii] Brooks (2011), pp. 45–46. The rest of the paragraph is a paraphrase of Brooks' comments.

[ix] Low (2002), Chapter 9.

[x] Cf. Sacks (1987), *The Twins*.

[xi] The dire results of society's overemphasis of analytic is the point of David Brooks recently article in The Atlantic.

[xii] All of the quotations in the sections on hemispheric differentiation come from McGilchrist (2009) where one may find additional references to the scientific literature.

[xiii] Marcel Kinsbourne quoted by McGilchrist.

[xiv] Religion might call this "original sin."

[xv] Cf. Low (2002).

[xvi] Cf. Davis and Hersh (1981).

[xvii] This discrepancy is not as great in science because the identification of a science with theory is not as complete as in mathematics. Nevertheless the same tension plays itself out.

[xviii] Wigner (1960).

[xix] McGilchrist, *op. cit.*

[xx] Cf. the article entitled, "Go Ahead Rationalize. Monkeys Do It Too." New York Times, Nov. 6, 2008.

Chapter 5

The Creative Process

5.1 Introduction

Creativity does not merely involve the production of novelty. It does not arise from applying the same procedure to a series of different situations. Creativity involves coming to see some situation or phenomenon in a substantially different way—it is not the phenomenon that has changed but rather the manner in which one views the phenomenon. Does the sun go around the earth or does the earth revolve around the sun? Going from one viewpoint to the other involves an act of creativity. In other words, creativity involves what I have been calling deep thinking. Deep thinking is the essence of the creative process, that is, the manner in which what I have been calling creative intelligence manifests itself in the human domain.

Since all deep thinking is creative, an investigation into deep thinking is necessarily an investigation into the creative process. There have been many such studies. What is different and unusual about this particular study? For one thing I regard creative intelligence, that metaphoric unity of creativity and intelligence, as something that is extremely general. Normally creativity is looked on as something that is the preserve of talented, highly intelligent people and is based on a particular way of using analytic intelligence. However, in my view, human creativity is a variety of something that is much more widespread and general. It is a capacity that can be seen in many animals, for example, in the way that crows and chimpanzees use tools and solve problems. Even more generally, the evolutionary process itself gives one the impression of sharing the characteristics of creative intelligence (as I mentioned in the preface). Evolution is punctuated by a series of discontinuities, which are very reminiscent of the kinds of leaps we have been looking at in development, learning, and paradigm change. We miss the intelligence of the process because we believe that evolution is "caused" by randomness and we take randomness to be the opposite of intelligence. We readily accept that evolution results in intelligence but we resist the idea that the process itself is intelligent. The progression of science in fields like cosmology and evolutionary theory has been notable for a movement away from human exceptionality. I suggest that creative intelligence and deep thinking are also very generalized phenomena, which we mistakenly identify with the purely mental phenomena that we observe in human beings.

The difference between my perspective and the usual one can be highlighted by asking the following question: "Is an act of creativity something that we do or something that happens to us?" In terms of the discussion in the last chapter this would be like asking whether creativity involves flashlight or lantern consciousness or whether it primarily involves the left or the right hemisphere.

Mathematics and science are notable for their remarkable creativity. The history of science is full of intellectual revolutions—the work of Einstein or Newton, for example—where some aspect of the natural world comes to be understood in an entirely new way. Thus after some preliminary words about creativity in general I shall turn to a discussion of creativity in science and mathematics. My aim is to show that some of the essential characteristics of creative activity are analogous to the features of deep thinking that I isolated in the earlier chapters.

5.2 What Creation Myths Tell Us About Creativity

What then is creativity? There is no question that is more mysterious or, in my opinion, more important. Even such basic questions as, "What is the essence of human nature?" "What is the origin and destiny of life?" and even "What is the good life?" need to be considered in the light of a proper assessment of the role and nature of creativity.

All cultures organize themselves around a story, which tells them how the world came into being—a creation myth. "In the beginning God created heaven and earth" is one such story. "The Big Bang" is another. Unfortunately we may misunderstand what these stories are trying to teach us. We may think, for example, that they happened at some time in the past but, of course, "in the beginning" there was no time. Time, as we understand it, is something that came into being.

If God created everything then this was an act of creativity. If the universe originated as a singularity in the space-time continuums then the Big Bang was an unparalleled instance of creativity in which something comes into being out of nothing. In either case creativity is intimately connected to the way we humans conceive of our origins. "In the beginning there was creativity," you might say.

Some people think that this kind of creativity happened a long time ago and that ever since then the universe goes on in an automatic way, like a machine, with no further need for additional acts of creativity. But there is another way to think about these things. It is that creation is ongoing—the world, the stars and the galaxies; ourselves, our bodies and our minds, are being created moment-by-moment, right now and will be forever. This ongoing process of creation is intelligent, that is creativity is intelligence in action. Contemplating this kind of fundamental creativity cannot fail to evoke feelings of awe and wonder. All creation stories—scientific, religious, or cultural—tell us in their own way that creativity is the primordial essence. Everything arises from this vast up welling of creativity and our own creative efforts come directly from that source—they have the same essence.

All human beings possess an inherent potential for creativity in just the same way as all children have the capacity to learn. Indeed learning and creativity are, in a fundamental way, different names for the same phenomenon. When we observe children at play we become aware that play is a manifestation of an inherent creativity and that we, too, are most alive when we allow our fundamental creativity to express itself.

We often mistakenly identify creativity with mere novelty, on the one hand, or the production of works of genius, on the other. We forget that creativity can also be thought of as an orientation towards the mundane activities of everyday life. Almost any activity—even doing the dishes—can be pursued in a creative manner; just stepping out of your front door can be a revelation. However those ordinary activities must be undertaken as though for the first time and not by habit or by applying some formula or set of rules. On the contrary, creativity means overcoming our habitual reliance on formulas and sets of rules. It requires opening ourselves up to the situation and experiencing it without preconceptions.

When we open ourselves up to what is real in this way what will we find? We sometimes imagine that the world is in some kind of permanent state that exists independent of us and is waiting for us when we step out the front door. We think of it as coherent and easily comprehensible with all the bits and pieces fitting together like the elements of a painting, a grand theory, or symphony. However it is conceivable that what is out there is dynamic; that the painting is constantly changing; that it isn't always coherent but that the patterns that we see regularly break down into incoherence and discord. A more poetic description of this dynamism was expressed by the Russian author Boris Pasternak in a letter to Stephen Spender: "I would pretend (metaphorically) to have seen nature and universe themselves not as a picture made or fastened on an immovable wall, but as a sort of painted canvas roof or curtain in the air, incessantly pulled and blown and flapped by a something of an immaterial unknown and unknowable wind."[i]

In fact the coherence that we see is the result of an act of creativity—the way that the world appears to us when we view it though a conceptual system. If this is so, then impermanence and incoherence are inevitability part of any situation. Incoherence may seem like a barrier but this incoherence may become a conduit towards creativity and greater understanding. The price of creativity would then involve facing up to the fear and anxiety that are evoked by the inevitability of impermanence and therefore the periodic breakdown of the mental systems that we have come to rely on. The grand creative breakthrough, as we shall see, arises out of confronting and passing through conflict and incoherence.

Talking about facing incoherence and passing through barriers may sound very dramatic, making it seem as though creativity is reserved for heroes. On the contrary, as I said earlier there is a possible creative act waiting for you every time you go out your front door. As you step outside on a beautiful spring day, the world seems to be reborn if you can only manage to see it with fresh eyes.

5.3 Creativity as Deep Thinking

An act of creativity is the result of an insight that arises discontinuously. The insight is often preceded by something that is deeply problematic; it is so deeply problematic that a resolution may well seem impossible. This is why the resolution does not arise through systematic means but only occurs when all systematic approaches have been exhausted to no effect, that is, if you want to be creative you must sometimes be prepared to fly blind. This is not easy to do. Creativity involves living for protracted periods with the kind of tension that arises in situations of cognitive dissonance.

Creativity needs to be distinguished from computation. Computational thought is used until it cannot be used any more—until it hits a barrier. Its importance for deep, creative thought lies in the anomalies that it uncovers. Of course these anomalies must be consciously accepted before further, creative progress can be achieved. That such acceptance is unusual and difficult will be seen in the remainder of the chapter. The problem with accepting the reality of anomalies follows from my discussion of hemispheric specialization in the previous chapter. Computation is *the* preferred mode of the left hemisphere and the left hemisphere is reluctant to accept anomalies or even to admit their existence. Its preference is for consistency and so it views new data through the lens of old schemas or, if that is not possible, it may deny outright the existence of the inconsistent data. Therefore creativity only can occur when the systematic mental picture of the situation breaks down and it is this breakdown that is experienced as painful. This explains the most important dynamic behind genius and creative breakthroughs. It takes an exceptional person to face up to the problematic; it is easier to just turn away.

The problematic element is not due to any mistake or error in the work of the creative individual but is intrinsic to any system of conceptual thought. We do not accomplish creative work by looking for a contradiction, let us say, in the current situation. It is that we become aware of the situation's intrinsic contradictions. We are caught in the following double bind: we cannot do science without some theoretical framework but every such framework is incomplete and contains problematic elements. We keep looking for the "truth," the ultimate theory, but such a theory does not and will never exist. The truth can never be fully contained within a given theory.

5.4 Creativity as Learning

One of the themes of this book is that learning in its most profound sense is creative activity. Child development can be thought of as a form of learning and so can the development of mathematics and science. Certainly in doing research we are motivated by the desire to learn about the phenomena that we are studying, that is, to understand it. As we shall see in the next chapter, there are many kinds of learning, including rote learning and algorithmic learning, but the most significant kind of learning, "deep learning" if you will, involves acts of creativity. Creativity is the common theme that ties together scientific research, child development, and deep learning. Creativity *is* learning in its most profound sense.

When I began thinking about these things I realized to my surprise that learning and child development had a great deal in common and were both forms of creative activity. This surprised me because learning seemed to be volitional and to require consciousness whereas development was built-in. If the same process was going on in both situations then maybe creativity is not a function of the analytic mind. Maybe it involves processes that lie below consciousness. It may even be true that normal, adult creativity is merely an aspect of something that is much more basic which I am calling creative intelligence. Perhaps the creative mind is no different than the "mind revealed in the world" as Einstein put it.[ii]

5.5 Creativity is the Correct Response to the Problematic

Creativity, like all deep thinking, involves overcoming a fundamental obstacle and so it is never easy. In this section I intend to discuss the response to that difficulty for this is the key to creativity. Yes there is something about the situation that is troublesome, that contradicts conventional wisdom, that may be irritating, and that produces tension. It is troubling yet the creative person cannot leave the situation alone, is unable to do the "normal" thing and just stop thinking about it. On the contrary, as Newton said, "… through concentration and dedication. I keep the subject constantly before me, till the first dawning opens little by little into the full and clear light." What Newton kept before him is not some well-formulated idea but a problem that is poorly formulated, full of ambiguity and contradiction. It is by allowing oneself to become totally absorbed in the situation with all of its problematic elements, living for protracted periods with the problem, and ultimately passing through the difficulty that the light finally appears. There is no path to creativity that avoids the problematic. This is a key element in creativity and it sets the tone for all that follows.

The Nobel Prize winning physicist Frank Wilczek in his book, *The Lightness of Being*, discusses the manner in which "a wrong definition can be the foundation for great scientific work." He goes on, "The legendary Danish physicist Niels Bohr distinguished two kinds of truths. An ordinary truth is a statement whose opposite is a falsehood. A profound truth is a statement whose opposite is also a profound truth."[iii] On the face of it this statement by Bohr is not logical but it is, in itself a deep truth and illustrates the distinction I made between the logic that applies within a conceptual system as compared to the creative logic that applies in going from one system to another. Bohr's comment is meant to apply to Quantum Mechanics and (possibly) to all of science but it extends, in my opinion to creative work in general. There are two kinds of logic (or ways of thinking) that are possible—classical or creative. The difference between the two is that the former can be considered as being "horizontal;" it is superficial in the literal sense of lying on the surface and of being consistent with earlier formulations of the situation. The latter is vertical; it is about change, often the radical change that completely reframes the given situation. In another book I characterized the difference between the two as "trivial" versus "deep," and as complicated versus complex.[iv]

5.6 Creative Logic: Holding Contradictory Thoughts in the Mind

For many people, and almost certainly for most scientists, "logic" means ordinary, everyday logic—there is only one proper use of the word. To such people "creative logic" could only mean obtaining creative results through the use of formal logical procedures. This would include the possibility of programming a computer to be creative on its own. This is not at all what I have in mind.

When I used the word logic earlier I was talking about ways of using the mind, of rational thought in its most general sense and which I will discuss at some length in a later chapter. It is a way of thinking. In other words, it might be better to speak of logical or algorithmic thinking versus creative or developmental thinking. Logical thinking is "easy" in the sense that it follows objective rules: for example, if **A** implies **B** and **B** implies **C** then **A** implies **C**. This is what makes it so attractive. Creative thinking is hard because these rules break down.

The writer F. Scott Fitzgerald wrote, "The test of a first-rate intelligence is the ability to hold two opposing ideas in mind at the same time and still retain the ability to function."[v] Here we are closer to a productive distinction between different ways of using the mind. Holding two opposing ideas in the mind at the same time produces cognitive dissonance and this may be unpleasant since it may create tension, discomfort, or even anxiety. What we habitually do in dissonant situations is make every effort to get rid of the dissonance. We do this by accepting one of the opposing ideas while rejecting the other—by calling one true and the other false, for example. We may even react by avoiding the entire situation—consciously or unconsciously erecting barriers that protect us from the conflict. Fitzgerald is suggesting that there is another possibility and that is to hold both ideas in the mind simultaneously even though it is difficult and unpleasant to do so.

Fitzgerald has put his finger on a key element in creativity. Creativity arises from the problematic, from dissonance, and logical contradiction is one form of dissonance. In *How Mathematicians Think* I spent considerable time discussing other varieties of dissonance like ambiguity and paradox. In whatever form it arises opposing ideas make us nervous and we crave consonance—a safe environment where things are predictable and under control. It is precisely this safety that we must give up if we wish to open ourselves up to a creative response. Can people be trained to work with situations of cognitive dissonance? This is a key question for educators at all levels and I shall return to it in subsequent chapters.

However a caution is in order here. It is that there is a part of a dissonant situation that is habitually hidden from view. One child says that there are no numbers between 1 and 2 while another, who accepts the world of fractions, says that there are many. These two answers seem to contradict one another, and they do, on one level. However on another level both children have an intuition of "number." Their different views arise because they view that intuition through different conceptual systems. There is a commonality behind their differences. Similarly the conflict of a football game is only possible because the two sides agree on the rules of the game. There is necessarily some commonality behind the emergence of conflict and dissonance. Thus creativity roughly involves the following dance: First, a situation with an implicit unity is interpreted in conflicting ways. Then the conflict is resolved and the unity is explicitly (re)-established.

In the next few sections I shall discuss the nature of creative activity in science as it is described by a number of writers and researchers. As we shall see, they all agree that creativity is difficult and discontinuous in the way that I have discussed. We shall also see that allowing oneself to enter situations of dissonance and remain there for prolonged periods is the key to the great leaps of science. Creative intelligence, what Fitzgerald calls a first-rate mind, lies more in this direction than it does in the direction of being able to make rapid calculations, store vast amounts of data in memory, or possess a very subtle and powerful critical faculty, which are the usual attributes that we think of when we speak of intelligence. Creativity is a response to the problematic.

5.7 Rothenberg and "Janusian" thought

Albert Rothenberg[vi] was a Professor of Psychiatry at Harvard Medical School who did a great deal of research on creativity in the arts and the sciences. He, too, considers a certain kind of incompatibility as the key to the creative process. To describe this incompatibility he coined the term "Janusian thinking," which he defines as "actively conceiving multiple opposites or antitheses simultaneously". (Janus was the Roman god with two faces, which looked simultaneously to the future and the past.) This sounds very similar to what Fitzgerald was talking about.

Rothenberg says, "During the course of the creative process, opposite or antithetical ideas, propositions, actions, or states are intentionally and consciously conceptualized side by side or as coexisting simultaneously. … Contradictory aspects are not reconciled but remain in conflict; opposites are combined, and oppositions are not resolved. Antitheses and opposites are held in tense apposition; they operate side by side, and in later phases, generate new and valuable constructions."

Rothenberg interviewed 22 Nobel laureates in the fields of chemistry, physics, and medicine and physiology. He also did extensive research on autobiographical accounts of the work of outstanding scientists of the past such as Einstein, Bohr, and Darwin. In most of the work that he studied he was able to identify the Janusian process as the essential element in the process of scientific discovery.

I do not have the space to give Rothenberg's work the complete description that it merits. However I will describe and comment on the four phases in the development of a creative breakthrough as Rothenberg sees them. The first is the motivation to create. This goes beyond a hunger for fame or mere curiosity. The scientists he studied were driven to be creative even if this involved difficulty and a kind of suffering. Einstein said, "The thought that one is dealing here with two fundamentally different cases, was, for me unbearable." And Hideki Yukawa who won the Nobel Prize in Physics in 1949 for his theory of mesons, said, "By confronting this difficult problem, I committed myself to long days of suffering." We have seen that this kind of difficulty characterizes all forms of creative thinking.

If creative work in science is so fraught with difficulty, why should one commit oneself to it? Think of the mathematician Andrew Wiles who spent seven years cutting off everything in his life that was not essential to his assault on Fermat's Last Theorem. Why make such a sacrifice for a result that is far from guaranteed? Of course we do not ask why a child is driven to stand up and walk. The child is responding to an imperative that it is powerless to ignore. The greatest artists and scientists also are powerless to fail to respond to the need to create. As difficult as creative activity may be, choosing not to do it is, for some, unthinkable. It would mean denying what is most important and alive in one's life. As the painter Van Gogh said of himself and his powerful need to create, "Something is alive in me: what can it be!"[vii]

"In Phase 2 of the Janusian process, one or both of the oppositional elements to be brought together are focused on or identified. It is here that the initial breaking away from the work of other scientists usually occurs." It is at this stage that the scientist has a hint, but only a hint, of the idea that will blossom into a new theory. Yukawa stated, "In retrospect, the basic concept of the meson came to me several times, but its appearance were [*sic*] like flashes of lightening that illuminated darkness only for an instant." Rothenberg maintains that, "Other scientists in this subject group have similarly experienced solutions and discoveries without full awareness or conviction."

"In Phase 3 of the Janusian process, pairs, sets, or series of opposites are simultaneously brought together in a conception that leads directly to the creative outcome. … the opposite or antithesis of an element in focus in Phase 2 may in Phase 3 be separately and substantively formulated; then opposites or antitheses are brought simultaneously together." For example, "From the formulation that an observer in free fall is both at rest and in motion at the same time, Einstein was able to postulate the relativity of motion…." Of course the paradoxical nature of complementarity in quantum mechanics was its most striking element. Bohr wrote, "This applies to every problem, that it is all about interaction between particles or between light and matter, and there we get at—according to nature—the irrational element of discontinuity-continuity or particle-wave. And in order to construct a complete formulation, inconceivable *means* are required."

None of Rothenberg's examples come from mathematics but there too we can see the same phenomenon at work. I have already discussed how the concept of number develops by reifying arithmetical operations like division or subtraction. In our normal way of thinking processes involve the movement or modification of objects. The objects are primary and the process secondary. These two are in essential opposition to one another and most languages make this explicit by using nouns for objects and verbs for processes. To think of a process as simultaneously being an object is difficult and initially strange. You get a hint of this strangeness when you consider the work of Cantor on infinity. As I discussed earlier "infinity as process" was acceptable but "infinity as object" was not. To think that an infinite process like the counting numbers could be considered one completed object is a leap.

To illustrate that this leap is still a difficulty for students (and many others) consider the following "paradox": "There is an infinite line of students in the hallway waiting to speak to Professor X. At one minute before noon the first two students, call them S1 and S2, enter the room but S1 leaves immediately. At a half-minute before noon, S3 and S4 enter the office and S2 leaves. At one-third of a minute before noon S5 and S6 enter and S3 leaves. This continues on and on. How many students are in Professor X's office at noon?"

As a process there are more and more students in the office at every interval in time: 1 at 1 minute before noon, 2 at a half-minute before noon, 3 at a third of a minute before noon, and so on. So the answer would appear to be that at noon there would be an infinite number of students in the office. But if you consider that at noon all of the intervals have passed and that everyone leaves at some moment in time there would seem to be no one left in the room at all. Do you think of this as a process that happens in time and therefore is without end or do you think of it as a completed process? This will determine your answer. Most students fall into the first category. They cannot hold these two seemingly contradictory thoughts about infinity in their minds at the same time. For them an infinite process must be incomplete and so can never be a completed object.

From the viewpoint of normal reason contradictions are unacceptable and must be eliminated. It might well seem irrational to say that a process is an object. After all an object is an object and not anything else and a process is a process. We desire a world that is intelligible to us, in which different words point to distinct features of reality; the language of science and prose in general we expect to be unambiguous unlike the metaphoric language of poetry whose essence is ambiguity. This is behind the Greek difficulty with zero and infinity and it is precisely what makes zero and infinity such creative breakthroughs. Both zero and infinity manage to put together opposites. I have discussed one set of antitheses contained in infinity[viii] but what are the antitheses in the conception of zero? For one thing, it is that the negative "zero as nothing" can be set against the positive "zero as the 'empty' source of all." The Greeks could not get to this stage but the Hindus did not have the same problem in formulating these opposites.

It is interesting that Rothenberg comments that, "I found them [the scientists he interviewed] somewhat reluctant to reveal the specifics of their thinking if it seemed, on the surface, irrational." All people share this reluctance and the more educated they are the more difficulty they seem to have with the non-rational (which seems to be a more suitable word than irrational). We have seen how this reluctance has its origins in our preference for observational flashlight consciousness and the left hemisphere over participatory lantern consciousness and the right hemisphere. This is why we are all more comfortable when we can situate ourselves within a mature theory, paradigm, or well-developed conceptual system. To admit the non-rational into our lives seems like a reversion to some primitive element in our nature. Even someone who is brave enough to descend into these depths and has returned, like the mythological hero, with a prize of great value still may be reluctant to talk about it and maybe even admit it to herself.

Now let me go on to Rothenberg's final phase, which involves the construction of the complete theory, discovery, or experiment. "It is in this construction phase that the structure of simultaneous opposition is modified, often to a point at which antitheses and apparent contradictions disappear completely. Some of the theories and discoveries retain a somewhat obverse or inverse quality and thereby give some hint of their development within the Janusian process, but, by and large, stepwise logic and external validation dictate their elaborated content. … In this construction phase, skill in mathematics and in deductive and inductive logic, as well as the creative scientist's comprehensive knowledge of a particular scientific field, predominate."

Logic predominates in this phase because ordinary logic represents the way the mind functions within conceptual systems especially within the systems of theoretical science and rigorous mathematics. We use the logical rules of inference to both organize our thoughts and to communicate them to others. More importantly logic integrates the new insights with the larger body of scientific or mathematical results.

At the very least subjecting your insights to the test of logical exposition may reveal holes in the argument. This is not just a verification phase but part and parcel of the creative process. It is in the stage of "writing things up" that the scientist or mathematician becomes aware of serious lapses that necessitate going back to the drawing board, to some extent at least. Thus the four stages in Rothenberg's model may be gone through more than once. In practice there may be many iterations of these stages; where lacunae in writing things up will lead the researcher to think things through some more and modify the original insight. This is exactly what happened to Wiles when his initial proof of Fermat needed to be substantially revised.

Finally it would be wise to remember that neither Rothenberg nor anyone else will give us ironclad rules for creativity. Nevertheless Rothenberg's fundamental insight is sound and consists of emphasizing the constructive role of contradiction within the creative process and thereby showing us that the rules that govern creative activity are not the rules of ordinary logic.

5.8 Koestler and "Bisociation"

Arthur Koestler's book, *The Act of Creation* appeared in 1964. In a preface he wrote, "The aim … is to show that certain basic principles operate throughout the whole organic hierarchy—from the fertilized egg to the fertile brain of the creative individual: and that phenomena analogous to creative originality can be found at all levels." Koestler's intuition, in other words, was that what I am calling in this book creative intelligence is extremely general and goes beyond the mental phenomena of certain exceptional people—that creativity is a property of all of organic life and, in particular, pervades all of human life. It is therefore instructive that he chose to begin his book with a discussion of humor, which he considered as a prime example of creative activity—not just making up and telling a joke but even "getting" a joke involves, for him, the essentials of creativity.

For Koestler the important thing was "the perceiving of a situation or idea ... in two self consistent but habitually incompatible frames of reference."[ix] He called this way of thinking, "bisociation", "in order to make a distinction between the routine skills of thinking on a single "plane" as it were, and the creative act, which, ... always operates on more than one plane."[x]

A good example of bisociation is the famous story of Archimedes who was asked by the king to determine whether a crown was made of pure gold. Obviously it was not practical to melt down the crown so Archimedes was stymied for a long time by the need to find the volume of the crown in another way. The resolution occurred when Archimedes stepped into a full bathtub and noticed that water spilled out. This was the second plane that led directly to the famous moment when, in his excitement, he reputedly ran down the street stark naked shouting, "eureka."

This incident also tells us other things about the act of creativity, namely, that it is sudden and can be precipitated by any accidental occurrence when the mind has been prepared through a lengthy period of absorption. This is serendipity, which is a normal feature of the creative process. Recall that the conceptual system defines what is real but that reality is always singular. Serendipity only happens when one becomes conscious of something new (the second framework). This happens through the right hemisphere and only occurs when the tendency of the left hemisphere to become fixated on a particular way of looking at a familiar situation is broken down.

Now Koestler's theory is not the same as Rothenberg's as the latter is at pains to emphasize in his writing.[xi] Nevertheless the parallels between these two approaches to creativity are instructive. Koestler's "self-consistent but habitually incompatible frames of reference" would correspond to Rothenberg's "actively conceiving multiple opposites or antitheses simultaneously." This "incompatibility" appeared in the conceptual development of children and was a feature of scientific and mathematical research. However in the latter situation it is preceded by an initial absorption in the problem and followed by the final stage in which the new theory or point of view is logically integrated with the larger body of scientific or mathematical theory.

The essential points brought out by both theories are that (1) creativity comes out of an incompatibility, (2) it involves a particular way of thinking or using the mind, (3) the ordinary rules of logical inference do not hold, (4) it involves multiple, inconsistent points of view and (5) a reframing of the original situation. There are striking parallels between both of these views of creative activity and the characteristics of moving from one conceptual system to another.

5.9 Martin and the Opposable Mind

Creativity is not only present in the arts and the sciences; it is also the holy grail of business and industry especially in recent years when young people can earn billions of dollars by inventing a bit of technology or new internet application. It is not surprising that the principles that I have been discussing should find their way into the world of business. Roger Martin, the dean at the Rotman School of Management at the University of Toronto has written about a method by which he claims we all can be more creative. In his interviews and study of leaders who have "striking and exemplary success records" he claims to have uncovered a common theme that runs through their successes. The people he studied have, in his words, "the predisposition and the capacity to hold two diametrically opposing ideas in their heads. And then, without panicking or simply settling for one idea or the other, they are able to produce a synthesis that is superior to either opposing idea. *Integrative thinking* is my term for this process—or more precisely this discipline of consideration and synthesis—that is the hallmark of exceptional businesses and the people who run them."[xii]

It is quite striking that we find Martin's discussion of creativity to be similar to the other authors that we have discussed in this chapter. Again there is mention of holding two contradictory or opposing ideas without flinching. Then, for Martin, there is the "integration" or "synthesis" of these ideas. The first would be a prerequisite to the second—what we normally call the creative breakthrough I believe that neither the word "integration" nor "synthesis" is well suited to this situation and shall go into this second phase of the creative activity later in the chapter. For now let's just note again that the setting for creative experience is allowing oneself to accept, without resistance, the entire situation complete with its non-logical elements.

5.10 Low and the Idea

The Montreal author Albert Low, in books like *Creating Consciousness* and the more recent *I Am Therefore I Think* sees the entire universe as the locus of a fundamental and ongoing creative process. His description of the creative process also breaks down into the phases that we have previously discussed and begins with the need to create which he calls "the yearning for unity." For him, this is the life force itself. The search for what is most alive in the artist, the musician, the poet, the scientist, or the ordinary person with a job and a family is the search for the creative. What comes out of this search—whether it is a song or a new scientific theory or a novel way of arranging the garden—is not as important as the process: the agony and the rewards of allowing the creative germ to express itself.

For Low the problematic element that we have been discussing lies at the origins of creativity (and equally at the origin of the human condition). In fact for him the problematic *must* be resolved—either constructively through an act of creativity or negatively through discord and violence. This violence need not only be directed at other people or at nature but may even be directed at oneself. The moral here is that the drive towards the creative cannot really be repressed; it must find expression. The only choice is whether that expression is to be constructive or destructive. Thus the decision to repress the creative has far-reaching implications. Not only does every person have a creative potential but denying that potential has far-reaching consequences for the individual's wellbeing.

Low goes on to discuss the suddenness of creativity. He, too, emphasizes the discontinuity of the creative insight. But what is the nature of this insight? Low talks about "the idea." For Low, the idea is not, and cannot be, explicitly formulated. Like a conceptual system, the idea is not something that we see but something through which we see. For that very reason we cannot see it. In the terminology of the previous chapter you could say that the idea is primarily a creature of the right hemisphere.

For example, "number" is an idea. You can't really define "number" although, of course, you can certainly define many different varieties of number such as counting numbers, rational numbers, real numbers, and so on. These, for Low, are concepts—things that have explicit definitions. But all these varieties of number arise from one generating idea, namely, number. You can say many things about number; you "know it when you see it" so it would be ridiculous to say that number does not exist. But it is limiting to pin it down. "Number" is dynamic because it is continually changing and evolving both within mathematics as a discipline as well as within the understanding of every student of mathematics. All of the basic ideas of science and mathematics are ideas in Low's sense.[xiii] A very partial list would include: space, time, energy, randomness, continuity, and so on. All are basic to a scientific description of the world. None can be pinned down definitively.

Idea is "that which reveals relations within phenomena." Low says, "Look around the room in which you are sitting. You see a chair: that is someone's idea." Everything you see is someone's idea. Ideas are the means through which we bring energy and coherence to our lives. They bring science into existence. In fact science itself is a magnificent idea that is still actively producing the modern world. Every mathematical proof, every scientific theory or experiment is the manifestation of a central idea. The idea is what is essential; it is the alpha and the omega.

Each of my books began with an idea, which appeared to me originally in a nebulous manner. Nevertheless this "feeling," this tentatively formulated idea, gave me the energy and enthusiasm to get going in a definite direction. Before the appearance of this central idea I had floundered for many months in a kind of wasteland, but then a chance remark or something I read in a book set off a veritable explosion and instantly I knew that there was an idea there. All of the hundreds of pages that were eventually written were implicit in that original moment. During the "wasteland period" I owe a great deal to friends who reassured me that this, too, was all part—an essential part—of the creative process.

The birth of the idea is expressed most poetically in the famous story of the great French writer Marcel Proust and the taste of the madeleine: "No sooner had the warm liquid mixed with the crumbs touched my palate than a shudder ran through me and I stopped, intent upon the extraordinary thing that was happening to me. An exquisite pleasure had invaded my senses, something isolated, detached, with no suggestion of its origin. And at once the vicissitudes of life had become indifferent to me, its disasters innocuous, its brevity illusory—this new sensation having had on me the effect which love has of filling me with a precious essence; or rather this essence was not in me it was me. ... Whence did it come? What did it mean? How could I seize and apprehend it? ... And suddenly the memory revealed itself."[xiv]

The taste of the cookie was the germ that led to an explosion, the idea, which opened out into the magnificent expanse of the many volumes of *In Search of Lost Time*.

5.11 Why There Can Be No Formula for Creativity

There are always people who purport to have a formula or technique for creativity. Sometimes they make a living out of teaching others their technique and there is a big market for such "teachers" especially in business where, as I noted earlier, innovation is a prerequisite for success. However, in this section I shall indicate why no such technique exists or can ever exist. This will teach us something essential about the nature of creativity.

The closest we can come to such formula comes from the thinkers that I have mentioned earlier. Recall the prescription of "holding two opposing ideas in the mind without flinching." Of course this doesn't tell us the nature of these opposing ideas nor their origins. It does not tell us when the reconciling idea will dawn or even if it will ever appear. It is, as we mathematicians want to say, a necessary but not sufficient prescription for creativity. The point here is that there are no sufficient conditions for creativity.

As I suggested in the preface, the sources of creative intelligence are not conscious. The "two opposing ideas" are conscious and so that is what we become aware of—that something doesn't work or is anomalous. Our beautiful, consistent system of thought breaks down and we are at our wits' end because the world of what we know can offer no resolution. That resolution, when and if it comes, can only arrive from nowhere, that is, from beyond our conscious experience. A genuine insight has this feeling of novelty. You do not know where it came from but you see instantaneously and with certainty that it will do the trick. That this is the right way to look at the problem.

Thus there can be no technique for creativity because any activity generated by a technique or formula is not creative. Algorithms, formulas, and techniques all lie on the surface. They involve analytic thinking, that is, using left-hemispheric processes. Creative thinking, in so far is it is original at all, necessitates the involvement of the right. As I said in the last chapter creativity begins in the right hemisphere and ultimately returns there. But the right hemisphere's contribution is not acknowledged by the left in which our sense of self is mostly lodged. It is, you might say, invisible to the left and another word for invisible is non-conscious. The entire process is that of going beyond—beyond the left hemisphere, beyond consciousness, beyond what you know and what makes you feel comfortable and secure.

5.12 Focused Thinking

The first stage in the creative process involves focused thinking, that is, flashlight consciousness, in which the problem is defined and its essential elements are isolated. This may lead directly to a solution if one can apply a well-known method to the situation. In this case, there would be no need to go further—no creativity is called for.

Normally when we discuss science and mathematics our conscious mental state and that of our audience is this focused consciousness, which, as we have seen in the last chapter, is the characteristic mode of functioning of the brain's left cerebral hemisphere. This leads to the analytic way of dealing with the world. Much of our adult lives are spent in this type of consciousness.

Before anyone can be creative in any field they must first master their craft and this involves a lot of hard and repetitious work—focused work. Many of us hope for creativity as a kind of liberation, which it undoubtedly can be. Unfortunately some people hope that the release will come without hard work—as a kind of grace. In other words, they want creativity without the kind of difficulty that I talked about earlier in the chapter.

If you want to be creative you have to pay the price and the price is high. The difficulties that are involved come in different forms. The first involves absorbing yourself with the problem and familiarizing yourself with the disciplinary language within which the given problem is formulated. I've certainly run into students who believed that creativity involved getting high and then just standing back and waiting for the miracle to happen. The moment of insight *is* a kind of miracle. You cannot force it to happen; you must allow it to come to you and this means acknowledging that "you" are not in control. I've even known some very successful and creative academics who smoke marijuana on a regular basis. Some of them claimed that it helped their creativity and perhaps it did and we'll see in the next section why this might be so. But what is certain is that these people work very hard at what they do—putting in endless hours in the lab or study, reading, thinking, and attending to the details of their work. Without this initial work and training nothing ever happens. Even if you have a brilliant idea, in order to communicate it to people you must speak their language. You cannot invent your own language and then wonder why you are not appreciated. At the very least you must be able to explain why your work is worth considering. All of this requires one to be in a state of focused consciousness.

However there is a second level of difficulty, one that I discussed in the previous pages, entering into a situation that is replete with confusion and contradiction and sticking with it without attempting to prematurely escape from the tension it generates within you by pushing it into some predetermined form. This is the time when logical rules break down and non-rational factors find their proper place. This is the crucible of creativity. Before new structures or ideas can come into being the old ones must give way. But they do not give way easily. We love our ideas and conceptual frameworks and resist letting them go. We hate cognitive dissonance and that is why a new idea takes so long to take hold. We hold on to our theories and ideas even in the face of direct evidence that they do not work.

Think of the instructive story of the winner of the 2011 Nobel Prize for Chemistry, Daniel Shechtman. He saw a crystal-like object with five-point symmetry in his electron microscope but theory said that crystals with such symmetry could not exist. However Shechtman tenaciously held on to his belief that what he saw was real in the face of almost universal skepticism. His research leader kicked him off his team claiming that what Shechtman saw could not exist because the textbooks said it was impossible. The great scientist Linus Pauling, referring to Shechtman, said that, "There are no quasi-crystals only quasi-scientists." We all desire the praise and approval of our peers. Most people would begin to doubt themselves in the face of such disapproval and ridicule. Shechtman is an intellectual hero because he *knew* that he had seen something significant and had the inner resources to pursue his vision come what may. All of this humiliation and ridicule indicates the tenacity with which we all, individually and collectively, hold on to our ideas and how reluctant we are to see them challenged much less changed.

This inertia has to be broken down if something new and significant is going to have a chance of emerging. This inertia results from inhabiting a fixed conceptual system together with its habitual way of thinking—logical analysis and focused awareness. It is precisely the problematic elements of the situation that will break the inertia of our habitual thought processes. This is why it is not possible to leverage pure logic into a new idea. There must be a problem and one must authentically live out the discomfort and tension that it produces. One must be prepared to sacrifice all of one's cherished preconceptions on the crucible of creativity and this includes the logical thought processes through which one structures one's inner world.

5.13 Letting Go

However just living in tension and incoherence is not enough. The creative idea is born out of intense absorption in the totality of the problematic situation. One works and works but eventually you take a break—go on vacation, read a book, take a walk, or just chat with your spouse and children. Archimedes took a bath; Poincaré went on a short trip; and it was then that the fateful idea burst forth. Why then and not when they were working so hard? The simple answer is that at that moment they allowed themselves to relax but the deeper answer is that "relaxing" means changing your mode of awareness.[xv] This means moving from the left hemisphere to the right or from focused consciousness to something more elementary, namely, Gopnik's "lantern consciousness," that is, from focused consciousness to an awareness that is unfocused and alert.

As I discussed in the last chapter, creativity requires the collaboration between mental processes that are housed in different hemispheres. It involves an alternation of left and right and so of flashlight and lantern consciousness.

Creativity involves a subtle use of awareness. Focusing takes an effort and an intention and we all have a feeling for what this means because focused awareness is our normal state of consciousness—the state that you are in when reading these words. Un-focusing is difficult to describe in words because when you are speaking or reading you are focused; the closest I can come is repeating the phrase "relaxed alertness" or "letting things be." Think of a cat waiting for the appearance of a mouse from a hole in the wall—totally relaxed but totally alert—let the mouse show even a whisker and it's all over.

Letting things be is often the most difficult thing for a conscientious person to do. It requires a kind of humility, a willingness to give up control. Remember the student who felt that he should be rewarded for the time that he had devoted to solving a problem. He didn't know when to quit even if he was essentially spinning his wheels most of the time. How many of us have had the same experience! When do you say to yourself, "I've made all the progress I am capable of at this time. I'm going to bed and will pick this up in the morning." If you can let things go and change gears you may find that when you sit down at your desk in the morning something has clarified itself in your mind overnight and a new approach to the problem occurs to you. Going to sleep is the easiest way for most people to alter their state of consciousness.

When I began to write books, this is one of the most important lessons that I learned. The process involved alternate periods of ordinary, focused work leading, often, to an obstacle which, for the moment, seemed insurmountable. The next step involved just getting up and changing gears, letting go, and coming back the next day. To my amazement and surprise I often found that the problem had resolved itself and all I needed to do was to write things down. The resolution felt miraculous because it seemed to come from nowhere and "nowhere" means that it did not arise from the analytic, focused mind.

The creative process is this dynamic situation that involves concentration and relaxation, periods of being in control alternating with periods of relinquishing control, the birth of coherence out of the problematic and the incoherent. It requires establishing a working relationship between focused and unfocused awareness. A prerequisite for creativity involves breaking our tendency to stay trapped within our present mental paradigm, our fixation, in particular, with the analytic, left-brained, way of viewing the situation.

5.14 Letting Go Is Impossible

But perhaps I am contradicting what I said about there being no formula for creativity by recommending my own technique—focus and then let go. However things are not so simple and the reason for this lies in the enigmatic expression "letting go." Letting go is not volitional. It is impossible to *decide to let go*. Even that formulation is not correct since you can indeed *decide* to let go but your decision will only be another block in the process of letting go. Nevertheless letting go can and does happen.

The injunction "let go," like the injunction "be spontaneous," is paradoxical in a way that reminds me of the "liar's paradox"—"this sentence is false." If you let go because of a decision to let go then you would not be letting go at all. The opposite of letting go is following some rule or formula, which is what you would be doing if you could decide to let go.

Letting go is not something that you can force yourself to do. Try it and see! However this does not mean that it is impossible to let go. Actually, if you try to let go and really think about what you are doing, you may well conclude that it is impossible. On the other hand, there is the undeniable fact that it sometimes happens—people manage to let go all the time. It's like the difference between the voluntary and involuntary nervous systems (somatic and autonomic). There are bodily functions that can be controlled and those that normally cannot. There are huge parts of our life, such as healing a wound for example, for which is it reasonable to say, "I can't make it happen, but it certainly does happen."

Creativity is this kind of paradoxical activity. If you could force yourself to be creative, if I could provide you will a formula guaranteed to produce creative results, then creativity would be the outcome of focused consciousness and the analytic mind. Being in control, following some algorithm or well-defined procedure involves looking at what we are doing from the outside yet, paradoxically, it always involves looking at things from within a well-defined system. It is from this point of view that letting go seems to be impossible. No one ever let go when they were in their normal mode of consciousness.

However, as I said above, creativity involves letting go and letting go involves changing from focused to unfocused awareness. Actually you cannot voluntarily decide to make this switch for exactly the same reason that you cannot decide to let go. Unfocused awareness *is* letting go. It is human beings' bedrock consciousness and the natural mind state of infants as we saw in Chapter 4. So much of the difficulty of creativity reduces to the question of how to move from one form of awareness to another. There are two basic kinds of consciousness or awareness—a discrete situation—this explains why creativity is sudden and discontinuous. At some point in the creative process you let go and every creative person will approach letting go in his or her own way. Let me just say again that letting go is not something you *do* or even decide to do. It is something that happens. It happens *to you* not *by you*. It cannot be programmed because it is does not obey these kinds of rules. Before it happens and you are observing the situation from the normal point of view, a resolution to the problem may well seem to be impossible. Afterwards it feels obvious and inevitable.

5.15 Conclusion

It is inevitable that we enter most situations from within our habitual mental framework, something like a paradigm or conceptual system. Then we meet some blockage, something that just makes no sense within the system. Our ability to stay absorbed in this problematic situation breaks down our mental set but does not replace it with anything. At this stage we are ripe for the creative idea to appear and anything can set it off. Whatever the crucial element turns out to be it is evident that it must come from outside of our present system. We can think of this crucial event as the larger world impinging on the limited world of our current way of thinking. That is why it seems to come from the outside and to arise accidentally. It is interesting that the theory of evolution also posits randomness as the precipitating factor. If learning is also a creative activity it explains the importance of play and exploration, which seem more suited to the production of new ways of seeing than formal, structured activities.

What are the consequences of looking at creativity in this way? What, in particular, are the consequences of my assertion that there is no possible algorithm, rule, or technique for creativity? If creativity cannot be produced by formula or technique then it must be a free creation of the mind—truly original with no direct line that connects it to what came before. This is what I have been trying to get at by the expression "creative intelligence," which I have been trying to say is an essential characteristic of mind, of life, and of the natural world. The evolution of life and the evolution of the cosmos consist of episodes of a most marvelous creativity that could not have been predicted *a priori*. Where did the Theory of Relativity come from? The transition from the inanimate to the animate? Beethoven's Ninth Symphony? In the face of these incredible leaps of creativity the idea that they could come out of an algorithm or a computer program seems downright silly. We cannot help but be inspired and reassured by the idea that the very same creative sources that produced all of these marvels are accessible to us all.

---

[i] Quoted in the introduction to *Doctor Zhivago* (2011).
[ii] Einstein (1930).
[iii] Wilczek, p. 11.
[iv] Byers (2007).
[v] F. Scott Fitzgerald (1945).
[vi] All of the quotations in this section are taken from Rothenberg (1996).
[vii] As quoted in Low (2002), p. 134.
[viii] A more complete description of this point will be found in Byers (2007).
[ix] Koestler (1964), p. 35.
[x] *Ibid.*, pp. 35–36.
[xi] Rothenberg (1996), p. 221.
[xii] Martin (2009), pp. 6–7.
[xiii] In *The Blind Spot* I called them proto-concepts in an attempt to get at the subtle difference between what can and cannot be made explicit.
[xiv] Proust (2003), Vol. 1, *Swann's Way*.
[xv] In the rest of the chapter I will sometimes use Low's terminology of awareness instead of consciousness for the reason that I mentioned earlier—consciousness is used in a transitive form, consciousness of something, and so-called lantern consciousness is intransitive. It is participatory rather than observational.

Chapter 6

The Cognitive Dimension of Mathematics

There is no logic and epistemology independent of psychology.
-H. Poincaré[i]

6.1 Introduction

From the beginning of the book I have been viewing mathematics through the lens of the cognitive structures that are used to create and understand it. Normally one thinks of the conceptual domains of mathematics as totally disjoint from the content. The content is thought of as primary and identical to the subject whereas the conceptual is secondary. This distinction between the content and the conceptual, though useful at times, is artificial. Ultimately it distorts our understanding of the subject and therefore produces all kinds of problems not only for individual learners but also for people who apply mathematics to their work and so to society in general.

The best way to think about mathematics is to include not only the content dimension of algorithms, procedures, theorems, and proofs but also the cognitive dimensions of learning, understanding, and creating mathematics. In an early critique of Formalism the great French mathematician Henri Poincaré wrote, "The logical point of view alone seems to interest [Hilbert].[ii] Being given a sequence of propositions, he finds that all follow logically from the first. With the foundation of this first proposition, with its psychological origin, he does not concern himself." I am interested in Poincaré's "psychological origin," that is, in the mind that produces and understands mathematics. The true foundations of mathematics do not lie in axioms, definitions, and logical inference, which are the foundational elements of formal mathematics. The true foundations of mathematics lie in the minds of mathematicians as they interact with and try to make sense of their world—in their ideas, their intuitions, and their aesthetic sensibility.

In fact a formal mathematical system is an abstraction of a basic cognitive structure—a conceptual system. The formal system is the social, consensual dimension of these conceptual systems with the cognitive elements stripped away. But it has no meaning, either mathematical or otherwise, unless it is brought back to life by learning, knowing, and understanding. Thus an abstract structure is not in itself mathematics. Learning and research are the processes whereby life is reintroduced to the subject. You make it back into real mathematics in the same way that you put life back into a research paper—not by reading it passively but by ferreting out its core ideas. If this process is successful it results in the exclamation, "Ah! Now I see what is going on." Real mathematics is alive and it is the mind that gives it life!

Viewing mathematics in such a dualistic way is the kind of situation that I have been calling a constructive ambiguity. You have one subject, mathematics, which can be looked at from two perspectives—the theoretical perspective of its content or the cognitive perspective of concepts and conceptual systems. These two perspectives are mutually enriching. The formal theorems and proofs deepen our understanding of the concepts involved. Conversely a deeper conceptual understanding leads to new and better theorems.

The theoretical and cognitive dimensions of mathematics are obviously not independent of one another. In this chapter and the next I intend to see what this correspondence can tell us about the nature of mathematics.

The deep relationship between content and cognition are not generally recognized because the dominant philosophies of mathematics, Platonism and Formalism, both suppress the cognitive dimension of the subject. They function as a conceptual system; in practice mathematics is constrained by what can be seen through the window of these philosophies. Platonism postulates some ideal domain that exists independent of human intervention in which the timeless truths of mathematics reside. Formalism explicitly identifies the theoretical dimension—the deductive system—with the entirety of the subject. I am not interested in the argument between Platonism and Formalism for the simple reason that these points of view assume that "mathematics" means the objective content of mathematics, its computational methods, theorems, and proofs. The time is ripe for another point of view—a new conceptual system—that is more consistent with the reality of contemporary mathematics.

Mathematical development begins at an extremely early age, and as we saw, it appears to be built up from core conceptual structures that infants are born with. Mathematical cognition is one of the most basic ways in which human beings interact with the world. We are mathematical beings. For what are no doubt good evolutionary reasons, nature has privileged the development of mathematical notions like number, order, and quantity. It follows that mathematics is a good place to look when we consider general epistemological questions. In other words, understanding mathematics properly will give us great insight into human nature, the structure of the natural world, and the relationship between them. Conversely looking at mathematics from a more general perspective will lead to a new way to understand and appreciate mathematics.

6.2 Real Mathematics is Conceptual Mathematics

Mathematics comes in two flavors. The first is the formal theory—Euclidean geometry is a good example—with its axioms, definitions, and proofs. The second involves human beings' internal representations of the subject—the cognitive dimension. The closest most people come to describing this dichotomy is when they discuss informal versus formal mathematics. "Informal" might refer to mathematics where one takes into consideration things like understanding and intuition and so the formal requirements—for proof, for example—are loosened. In recent years we have seen more of mathematics in which the formal requirements have been relaxed somewhat—experimental mathematics,[iii] for example, which is associated with the sophisticated use of computers. Even here there is a debate about whether this kind of work is "real" mathematics, where "real" refers to the classical ideal of mathematical rigor. However the consensus view about what constitutes real mathematics is something that has changed many times in the past and is changing today.

We stand today at the end of an era which is notable for an attempt to make rigor and formalism into the defining characteristics of mathematics. This attempt arose out of the crisis associated with the advent of non-Euclidean geometries and the problems and paradoxes associated with calculus and the real number system. It was an attempt to make mathematics completely objective and universal, which in practice meant separating the content of mathematics from the process of doing mathematics—mathematics as knowledge from mathematics as a way of knowing.

You can see the beginnings of this tendency as far back as Euclidean geometry when Euclid tried to supply a precise definition for everything that formed a part of his system. Unfortunately it turns out that you just cannot define everything and so some of his definitions—of points and lines, for example—look contrived and circular to the modern mathematician.[iv] Euclid's arguments look as though they are logically solid but they are usually made with a geometric diagram in mind. From one point of view this is a weakness of his system because the arguments often depend subtly on the diagram. However you can also think of this "deficiency" in the opposite way, as a strength of Euclidean geometry. Yes, it is true that we bring our geometric intuitions to Euclidean geometry. This does not demean this brilliant attempt at creating a system that is deductive and rigorous. On the contrary, it tells us something valuable about the proper role and value of rigor in mathematics. Euclidean geometry is a wonderful subject, not because it does away with the human element, but because it so successfully marries logic and intuition, simultaneously building up a deductive system while it engages and trains the creative resources of the practitioner. The notion that the role of a deductive system is to do away with the cognitive element is a mistake that arose with the excesses of formalism. Euclidean geometry tells us that logical rigor can coexist with the deepest intuition. The resolution of almost every deduction and proposition in Euclidean geometry requires a spark of creativity—often a geometric construction. Generations of mathematicians, scientists, and even people in the humanities have found Euclidean geometry to be a training ground, not only for logical argumentation, but also for creativity and deep thinking.

Let us admit openly that there is no "real" mathematics if by that we mean a subject that is objective and will never change. Let us give up on the simplistic version of both the Platonic dream and also the dream of Formalism. Real mathematics is conceptual mathematics. Real mathematics is a creative and developmental art form. The concepts and ideas that are its essential elements do not live in the textbook or research paper or in formal mathematics of any sort—these things are merely representations of mathematics and their principal purpose consists in evoking living ideas and concepts. Ideas and concepts do not exist in print; they exist in the minds of mathematicians. This is probably what the great French mathematician René Thom had in mind when he said that a theorem is true if the five best mathematicians in the field affirm that it is. Concepts live within the mind of the individual but they also live in a social context—in the classroom and in the community of mathematicians.

6.3 Consequences of Looking at Mathematics in a New Way

What are the consequences of accepting the idea that real mathematics is conceptual mathematics? For one thing we would have to look at the problematic aspects of mathematics and of learning in a different way. The growth of the subject takes place in the minds of individuals and the mathematical community as a whole. It has its origins in the places where problems and breakdowns occur. To say, "This does not make sense," or "This is a contradiction," and leave it at that will not get you anywhere. Why does it not make sense? How can I change the situation so that it now makes sense? These are better kinds of questions. All systems have their limitations, their regions of validity, and interesting things happen at the boundary. An interesting paradox is worth its weight in gold—you can be sure that there is a lot of good mathematics that is hiding inside of it. Think of the wonderful example of the Cantor set, which is tiny because it has length zero but is huge because it has the same cardinality as the real numbers. Enormous mathematical riches came out of this example: not only a deeper appreciation of the real numbers but also an introduction to deep mathematical ideas like strange attractors and fractals.

Advances in mathematics, and advances in a person's understanding of mathematics, come out of the resolution of problems. This does not happen if non-logical elements are avoided like the plague. It equally does not happen if there is no place for informal, intuitive approaches. There is a major difference between an idea and the formalization and logical derivation of that idea from a set of axioms. Mathematics is ultimately about ideas and ideas have a cognitive dimension.

6.4 Mathematics is Informal

Shifting our perspective on the nature of mathematics would change the relationship between formal and informal mathematics. We saw in Chapter 5 that rigor and formalism only come in at the final stage of the creative process whereas the other stages are informal. We are used to identifying the entire subject with its formalized version, which is what we see in research monographs and often in textbooks. From a formal perspective the informal is not significant—the result is all that matters; how you get there is not as important. A cognitive approach would change the relationship between the formal and the informal—how you get there—your understanding—is every bit as important as the final result. The formal without the informal is sterile.

Informal mathematics lacks the complete rigor and precision that come from being embedded into a strict deductive system. Children's mathematics is informal but it would be a grave mistake to regard the informal as a childish perspective that we should attempt to grow out of. A better perspective is that all of mathematics, especially the process of doing and learning mathematics, is informal. The formal is best viewed as an element of the informal that interacts with other elements. For example, you are studying some part of mathematics and you get an idea of what is going on by working with examples, with geometric diagrams, and so forth. This is informal work. Then you try to write it up (formalize) and realize that there is something wrong with the initial idea. Out of this comes another idea (informal) that you try to incorporate into the proof (formal) and so on.

Historically, a great deal of mathematics was informal if judged by modern standards. The work of Euler on infinite series was not completely rigorous. The development of the calculus by Newton and Leibniz needed nearly a century of work on its foundations before it could be developed in a manner that the modern mathematician would consider acceptable. Even the geometry of Euclid was not rigorous. Yet Euler, Newton, and Leibniz—not to speak of Euclid—were some of the greatest mathematicians of all time. Obviously these people were doing mathematics and clearly logical argumentation was part, but not all, of their work. However the use of logic is much more subtle than most people today give it credit for. The informal methods that were in use for calculus and infinite series were accepted until they began to produce anomalies and paradoxes. Then it became necessary to delve more deeply into the foundations of the subject in order to sort out what was really going on. Again the process is driven by the problematic. The logical exposition arises in response to the problematic but it also generates additional problems and insights. In other words, looking at things in this way integrates the process of rigor within the larger and more fundamental process of doing and learning mathematics.

The whole situation is beautifully illustrated by the case of the great Indian mathematician Ramanujan[v] who came up with the most extraordinary mathematics with only a rudimentary knowledge of the nature of proof. His mentor, G. H. Hardy at Cambridge, had a great deal of trouble trying to explain to him the relatively new European standards for rigor and convince Ramanujan of its necessity. One might say that most, if not all, of Ramanujan's mathematics was informal but no one would deny that it was mathematics at the very highest level. Ramanujan—the man and the work—beautifully illustrates the mysterious depths of mathematics and the creative process as it arises in mathematics. It has a great deal to teach us about the nature of mathematical activity precisely because one of the seemingly indispensible elements—its formal structure—is present in only rudimentary terms.

A recent paper by Reuben Hersh[vi] illustrates the uneasy relationship between formal and informal mathematics, and pure and applied mathematics. He discusses the paradoxical manner in which infinite series are dealt with in modern mathematics. The essential question about infinite series is whether they converge (add up to some definite number) or diverge (do not converge). According to theory, whether a series converges or diverges only depends on the "infinite tail" of the series—the initial terms, a_1, a_2, …, a_N (where N can be very large), can be changed or omitted without affecting the convergence of the series. On the other hand, suppose you are given some actual series and someone asks you whether you think it converges or diverges. What do you do in practice? What would an applied mathematician do? Hersh points out that they do what every student would do—they start adding, perhaps with the aid of a computer. If they discover that the resulting sums are getting closer and closer together—say after adding the first five terms all subsequent sums agree to two decimals places, after 20 terms they all agree to 5 decimals places, after 50 terms to 15 decimals places—then most people will conclude that the series converges. Notice that all of this information is obtained by using a finite number of terms, which the theory says are irrelevant. This process is of no formal value whatsoever but it is of great heuristic or informal value. Only after you have very good informal evidence that the series converges would you try to prove it.

Informal mathematics often works in this way—by induction and through the recognition of patterns. Of course some people are more sensitive and see patterns where others see nothing. Ramanujan is a famous example of someone who discerned regularities that no one else even dreamed of. Someone commented about his formula for the partition function that it was extraordinary that anyone intuited that such a formula might exist, much less the particular form that the formula would take. We tend to think that Ramanujan was so creative despite his lack of formal training but maybe it was precisely that lack of training (and also the particular form of his religious beliefs) that made him such an original mind.

As I have pointed out throughout the book, in order to do creative work the mind must go through alternating periods of focusing and relaxation, the dominance of right-hemispheric processes must alternate with those of the left. Rigor is connected with the left hemisphere: with the focused mind, analysis, and with closing down in a certain way. The informal is connected to the right hemisphere: to intuition, synthesis, mathematical play, experimentation, and opening up. Mathematics involves an interaction between the two. Most people give the last word to the analytic and formal but I prefer to give it to the playful and the creative, that is, the informal.

6.5 Concept and Definition

Another thing that comes out of questioning the role of the logical structure of mathematics concerns the relationship between concept and definition. In formal mathematics a concept is identical to its definition. Every element in the (formal) mathematical universe has a definition and conversely if it does not have a strict definition it is not a part of mathematics. In this vision mathematics is about precisely defined entities, their properties and relationships. One begins with the definition and then develops the properties of the defined object and its relationships with other well-defined objects.

Does informal mathematics use concepts, and if so, what kind of concepts would these be? When a developmental psychologist talks about "core concepts" or a baby's "concept of number," are they using the word "concept" in the same way as a mathematician does when she discusses continuity? To answer such questions necessitates a discussion of the nature of concepts.

What is a "concept?" This is a question with a long and distinguished history that I don't have the space to go into in the depth that it deserves. Many eminent thinkers have weighed in on this question, and not surprisingly, they do not agree with one another. I shall try to isolate some of the salient questions in the next couple of paragraphs. However first let me own up to the fact that I have been using the word "concept" in an informal manner that my usage has not been completely consistent.

What is the difference between the use of "concept" in the work of mathematicians and the "concept" used by developmental psychologists? For one thing the mathematician's concepts are based on words and symbols whereas the baby is non-verbal. There is indeed a debate about whether concepts depend on words and there are very smart people on both sides of this debate. The linguist Noam Chomsky, for example, argued for the verbal dependence of concepts whereas psychologists such as Jean Piaget claimed the opposite. Most mathematicians and scientists would probably say that concepts depend not only on words but also on logic and symbolic representations.

A second question is whether concepts are, in the words of the great philosopher Emmanuel Kant, *a priori* or *a posteriori*. That is, are they built-in or do they come from experience? The notion of core concepts seems to lean in the *a priori* direction. On the other hand, there is the view that a concept is abstracted from a multitude of specific experiences. Let's reconsider the concept of "number." It is never defined in mathematics even though it is used all the time and could be considered the most basic element in all of mathematics. What *is* defined are the various number systems that I have spoken of—counting numbers, rational numbers, real and complex numbers, and so on. Now you could consider "number" as an abstraction of these various specific kinds of numbers. The problem with that approach is that this would mean going in the opposite direction from the developmental and the historical. We "know" something of the concept of number before we know any numbers beyond one, two, and three. In fact we have (at least) two core systems devoted to "number."

Some people, bearing in mind the distinction I made between formal concepts and their more informal variety, have substituted the word "numerosity," for the word "number." Numerosity refers to the "tendency towards number." I used the term "proto-concept" in my book *The Blind Spot*, for the same reason, namely to distinguish between an informal tendency and the formally defined entity.

It would seem that we are born with a tendency towards ordering the world by means of "number" and the related idea of "quantity." These early core systems give us a direction for further development. As I mentioned earlier these core systems, and therefore this tendency, remain active throughout our lives. The tendency remains active because the fundamental problem, "What is number?" or "What is quantity?" or more simply, "How much?" and "How many?" is never fully resolved. Through our experience of number at home and at school we develop more and more sophisticated answers to these questions in the form of specific number systems: the counting number, the integers, the fractions, and so on. Experience and innate tendencies combine to produce a concept of number that is developmental—that changes and grows—and contains elements of all of the six categories that I have been discussing: formal, informal, verbal, non-verbal, *a priori*, *a posteriori*.

Kant also used the categories of analytic and synthetic to refer to concepts and I have discussed this distinction elsewhere in the book. This gives eight inter-related properties associated to the idea of "concept." It may well be that a concept is a complex mélange of all of these properties. It has verbal and non-verbal aspects. It arises from something that is built-in but may also be abstracted from many specific experiences. It has both an analytic and a synthetic dimension. All of these categories should not be approached as mutually exclusive but rather as illustrating different dimensions of "concept." If we ask the question, "What is (the concept of) number?" we should not be expecting a "correct" and definitive answer but a process that grows and changes in time. A concept is not a fixed entity.

Concepts have a "denoted meaning" and a "connoted meaning." The connotation is something like the "informal" meaning. Recall Tall and Vinner's "concept image," the totality of all possible ways of thinking of, or working with, the concept. The connoted image is not "well-defined"—you cannot look it up in the dictionary. It is malleable and can always expand to cover new territory; it can always be viewed in a new way. Every time a poet writes down a metaphor—something of the form **A** is **B**—she is giving us a new connotation for **A**. Even scientific ideas like time, space, and number have a metaphoric dimension. Actually they *are* metaphors as opposed to being objects but this does not make them unreal or vague. The richness of and multi-dimensional nature of concepts is highlighted by David Tall's use of the evocative term "crystalline concept" in his recent book *How Humans Learn to Think Mathematically*.

However a concept is not merely a collection of disparate metaphoric connotations because a concept is singular. It may be hard to specify exactly what we mean by "number" but we have the feeling that there is indeed some particular thing to which "number" refers. The elements of the concept image are focused or synthesized. This tells us that the connoted meaning of a concept is a matter of the synthetic intelligence which you will recall functions by creating unities or gestalts—one pen, one desk, one idea for "number" or "space" or "time." One unity that can be looked at from multiple points of view.

The denoted meaning, on the other hand, is clearly analytic. In science it is generally assumed that we are dealing with "denoted concepts" with precise definitions. But even in science this is not true. Consider concepts like time, space, number, randomness, and so on. We "know" what they mean; we have an intuitive feeling for them, even if we are not able to produce a definition. This tells us that we are dealing with the "connoted meaning." We may then try to make the meaning precise. We may postulate, for example, that space is a three-dimensional continuum consisting of all possible ordered triples (x, y, z), where x, y, and z are real numbers. This will capture some but not all of what we mean when we use "space" in a more informal, connotative sense. Nothing precludes a connoted concept from generating multiple denoted meanings. For example, we have the flat Newtonian three-dimensional space (Euclidean space) versus the four-dimensional curved space-time of Einstein. Different denoted meanings can even generate theories that are disjoint as we find with the concept of randomness, which can be developed into the normal theories of probability and statistics or into the algorithmic randomness of Kolmogorov and Chaitin.

Attempts to make the meaning of a concept precise are attempts to pin down the synthetic, connoted meaning and are always tentative and necessarily incomplete because you can't "capture" a synthetic and informal idea in the same way that the left hemisphere cannot capture the activities of the right. Capturing it would mean transforming it into an object that is suitable for use by the analytic intelligence. It means transforming the concept from one cognitive domain to another. The analytic and the synthetic each have their own strengths and weaknesses.

Science and mathematics properly involve the relationship between the two, not the attempt to reduce one to the other.

This is a constraint that science and mathematics will always have to work with. The dilemma is that we must try to make things precise but the very terms that we use cannot be made (completely) precise. This can be thought of in either a positive or negative way. The positive implications are more acceptable in mathematics than in science. They are that mathematics and science are processes without end, in the same way that music and art have no end. Science and mathematics will continue to produce an endless stream of new meanings for basic ideas and new creative insights into the nature of the physical, biological, and mathematical worlds.

6.6 The Abstract and the Concrete

Abstraction is an essential element of concept formation and mathematics is the discipline that has investigated the process of abstraction in greater depth than anywhere else. I have introduced the way mathematics handles abstraction in Chapter 3 but this may be the appropriate place to say something more comprehensive about the abstract and the concrete as categories of mathematical cognition, as opposed to the usual way of looking at them as categories of mathematical content. Abstraction makes most sense when it is looked at from a cognitive point of view.

Mathematics is simultaneously the most abstract and the most concrete of disciplines. We know it in a concrete and most intimate way because it is implicated in a number of our core conceptual systems. The counting numbers are almost tangible to most people, and at this level, it does not feel as though we add number to a pre-existing world so much as that we see the world through the filter of number. Babies can discern quantity before they learn to speak and so you could claim that "number" is more elementary than "word." Number structures the natural world and our own inner cognitive world, and so it is the most basic bridge between the two. If anything at all is concrete it is the core ideas for number that I discussed in Chapter 1.

On the other hand, mathematical structures are among the most complex and abstract creations of human intelligence. It is extraordinary that human beings have managed to create systems as subtle and abstract as the real and complex numbers, fractals, or the proof of Fermat's Last Theorem. They are the human creations that best approximate the incredible complexity of the human brain or the structure of DNA. One of the deepest tendencies of the natural world seems to be the tendency towards structured and hierarchical complexity and this tendency is mirrored in the cultural world of human creativity—in the great novels of Tolstoy, the plays of Shakespeare, the symphonies of Beethoven and Mahler, in the physics of Relativity and Quantum Mechanics, and in the sweeping structures of pure mathematics.

There is a curious relationship between the concrete and the abstract, the empirical and the theoretical. Does the concrete necessarily presuppose the abstract, or vice versa, do we need abstract categories to perceive the concrete? Do we need the concept of number in order to conceive of the counting numbers or is "number" an abstraction of our experience of working with the small positive integers? In general, is the world we sense a concrete manifestation of general laws? This is the way we often think about science: there is the law of gravity that determines the trajectory of the apple that falls from the tree. People once thought—and many people still do—that the universe comes with a set of blueprints—abstract laws—which determine a great deal of the subsequent history of the cosmos. This attitude is behind the old expression, "God is a mathematician."

On the other hand, there is a common belief that the abstract arises out of the concrete. You have a mass of data. You abstract the data by ignoring certain variables and measuring others but also by describing the data in terms of certain concepts—the falling apple has mass, position, velocity, and so on. Then if you observe patterns or regularities you may give these names and call them "laws," like the law of gravity. It seems almost obvious that the abstract arises from the concrete and is a step away from the reality of immediate experience. Yet the abstract principles often seem more real than the experience. Would we even be aware of the experience if we had no conceptual apparatus with which to describe it? The answers to such questions are not at all obvious and remain controversial. In this section I am interested in the relationship of the abstract to the concrete in mathematics and what this tells us about the general relationship between the concrete and the abstract.

In the 1960s there was a movement in mathematics education called "the new mathematics." It attempted to teach mathematics on the basis of certain abstract principles. For example, for every pair of numbers a and b, $a + b = b + a$ and $ab = ba$. These relationships are called the "commutative laws" and so students were taught to repeat that $3 + 11 = 11 + 3$ *because* of the commutative law. It was felt that these laws explained the reason behind certain observed regularities of arithmetic. But does a "law" explain why something is true, why something happens? Does an apple fall from the tree "because" of the law of gravity?

The "new mathematics" was an attempt to put into school mathematics a philosophy of mathematics that was dominant at that time in the world of research. Much of modern mathematics involves the study of structures that are defined abstractly, in terms of a finite list of properties. Think of the integers under the operation of addition. Its most obvious properties include: (1) when you add two integers you get another, (2) the order of addition does not matter, (3) when you add zero to any number the number is unchanged, (4) every integer, a, has a negative, $-a$, such that $a + (-a) = 0$, and (5) for any three elements, a, b, and c, $a + (b + c) = (a + b) + c$. If you begin by postulating that a set of mathematical objects have these properties of the integers (and only these properties) you get what is called an "Abelian or commutative group." Mathematicians have developed an elaborate theory of such groups studied by "algebraists" or "group theorists." This is the process of abstraction and it is a very powerful way of thinking.

However development and learning work in the opposite direction. The commutative law is not true for children *a priori*. It is learned as a result of a great deal of experience working with positive integers, most notably when the child sees that the most efficient way of adding $3 + 12$ is to start with the 12 and count 3 more instead of starting with the 3 and counting 12 more. Thus learning usually begins with the concrete and proceeds to the abstract.

I mentioned earlier that the system of real numbers is very complex and difficult. However we can study it more efficiently by abstracting certain of its more obvious properties, like the fact that its elements form an Abelian group under addition and that its non-zero elements also form such a group under multiplication. Other properties that have been abstracted from the real numbers and turned into mathematical subjects to be studied in their own right include the fact that there is a natural order for the real numbers which is preserved under addition and multiplication by positive numbers. Also there is a natural distance (or metric) between two numbers and this also can be abstracted and studied as an abstract structure in its own right. Each of these abstract structures that arise out of the real numbers is a flashlight that illuminates different important aspects of this number system. This process highlights the value of abstraction, namely, its role in focusing the attention on particular aspects of phenomena and ignoring others.

Since abstraction focuses attention on some particular set of features, it follows that it uses flashlight consciousness and the analytic intelligence. Abstraction highlights aspects of the environment that are of concern and ignores everything else. In this way the relationship between the abstract and the concrete has its roots in the structure of our brains and our consciousness. Abstraction is a natural use of the analytic intelligence just as the concrete and the immediate have an affinity to the synthetic way of using the mind. We naturally go back and forth between the two.

6.7 Is Mathematics Logical?

One of the most important consequences of thinking of the cognitive dimension as an intrinsic part of the discipline of mathematics concerns the role of logic. For mathematicians and non-mathematicians alike, mathematics and logic are inseparable. Mathematics is logical, and conversely, formal logic is mathematical, that is, mathematical logic is today a well-established field of mathematics. Mathematics and logic are so enmeshed with one another that many people would find my assertion that there are aspects of mathematics that are non-logical to be surprising. Nevertheless I believe that an effort to establish the appropriate role of logical argumentation in mathematics is important not just for mathematics but for the sciences and all other disciplines that use mathematics on a regular basis. It will send a message to all those in whatever field of human endeavor who use logical argumentation as the ultimate arbiter of truth.

Logic is a way of using the mind; it is a formalization of the process of thinking and reasoning. This whole book was an attempt to establish the claim that there is another way of using the mind, what I have called creative intelligence or deep thinking, and that this other way of thinking is to be found in mathematics. Let me recall a few elements of my discussion in Chapters 4 and 5 that bears on these questions. I distinguished analytic thinking, which uses ordinary logic, from synthetic thinking, which does not. I argued that both had an important role in mathematics. Also I maintained that deep thinking, the thinking that is involved when you move from one conceptual system to another, is not strictly logical, or if you will, another kind of logic is involved that I called developmental or creative logic. In Rothenberg's discussion of creativity in science we saw that formal logic had its proper place in what Rothenberg designated as the fourth stage of creativity. The earlier stages were not only non-logical but also involved the systematic use of contradiction. In my earlier writing I also pointed to constructive uses of ambiguity and paradox in the historical development of mathematics. It follows that to discuss mathematics from a developmental point of view, we must reassess the role of logical inference.[vii]

The discussion about the role of logic is necessarily subtle because most of us are much more familiar and comfortable with the analytic than the synthetic. In our stance as observers of the world, which is the stance we take when we discuss science, we are well on the way to the analytic. We observe the world, break it into pieces, and reassemble those pieces according to the laws of science and the rules of logic. This means that it is difficult to observe the synthetic much less to discuss it. A book, a research paper, or a theory cannot help but privilege the analytic since its final form comes from using that particular way of thinking.

For similar reasons it is difficult to get one's hands on the non-logical elements in creative thinking. The last stage in scientific creativity involves "writing things up" and this means translating the result into a logical format. This is true of all of science and not just of mathematics—all scientific journals have very strict requirements about the style of discourse and the criteria that are used to justify the "truth" of the result. This amounts to saying that the non-logical elements that are involved in the process of scientific discovery are inevitably hidden from view. Almost every journal article in mathematics or science cannot just be read but must be deciphered. The author has removed the living informal subject and replaced it with a formal and rigorous argument. The reader who hopes to learn something from the paper must take the formal presentation and recreate in his or her own mind that living essence that has been suppressed. That is why research papers in mathematics are hard to read unless you are deeply involved in the field. This process of moving from the informal to the logical and back again tells us something important about the nature of science and mathematics.

Logic—classical predicate calculus—and mathematics are where we look to establish the objective truth not only of science but also of the social sciences, of philosophy, and ultimately to justify decision-making in government and business. When I question the assumption that logic is *a priori*, that it stands outside of human culture and so is an objective guide to the truth, that it is the only correct and productive way of thinking, and that all other ways of using the mind are, at least potentially, reducible to formal logic, then I am taking issue with one of the axioms of a good part of Western culture. Educated people are so dependent on using logical inference to orient and structure their lives that questioning logic as an ultimate value can seem like questioning rationality itself and therefore may have the psychological effect of bringing on a kind of vertigo, a feeling that one has lost one's footing, a kind of descent into a form of chaos. We should take note of this feeling if it arises because this is the kind of ferment out of which creativity arises. It is exactly the state that we implicitly urge students to enter when we encourage them to learn something new.

6.8 Predicate Calculus

Let's take a moment and consider what we mean by logical argument in mathematics. Mathematicians do not use logical truth tables to structure their arguments but they constantly use a few informal logical rules, which I now list.

(1) The most important of these would be the rule of implication: if **P** and **Q** are two propositions, and **P** implies **Q**, then if **P** is true **Q** must also be true. Since every mathematical proposition is an implication consisting of a **P** (the hypothesis) and a **Q** (the conclusion) this logical rule allows one to build up linear strings of "truths": **P** implies **Q** implies **R**, and so on. Here are two examples: (a) If m^2 is an odd integer then m is also odd; (b) If T is a triangle in the plane and the axioms of Euclidean geometry are true (especially the parallel postulate) then the sum of the interior angles is equal to two right angles.

(2) The logical rule that enables you to turn an implication "**P** implies **Q**" into the (logically) equivalent implication "**not P** implies **not Q**," called the contrapositive. For example (a) above, this would be "if m is even then m^2 is even." For example (b), if the sum of the interior angles of T does not add up to two right angles then one of the axioms of Euclid must be invalid.

Logically a statement and its contrapositive formulation have the same truth status. The latter is sometimes easier to prove (directly) than the former (as in example (a)) so that their status as mathematical statements (as opposed to purely logical statements) is not identical.

(3) "Proof by contradiction": Instead of assuming **P** is true and directly trying to prove that **Q** is true, one assumes **P** is true but **Q** is false and deduces that some other well-known, true mathematical statement, like $1 \neq 0$, is contradicted. The Greeks liked this method of proof and used it to show that the square root of two was irrational and that there are an infinite number of primes.

(1), (2), and (3) are the main logical rules of the working mathematician but there are two other logical rules that are important in mathematics but are not always made explicit.

(4) The "law of identity" that ensures that mathematical concepts remain constant. (In Gertrude Stein's phrase, "A rose is a rose is a rose.")

Thus π always refers to the same constant and an equilateral triangle always has the same definition. Another way of saying this is that (formal) mathematical concepts are pinned down by their formal definitions. A "prime number" is always an integer with no divisors except itself and 1. What it means for a function to be continuous is totally captured by its definition—a complex logical statement that I will not bother with here. This constancy is true for the elements of theoretical mathematics and science but not for their conceptual elements. A concept is not identical to its formal definition as I discussed earlier. Concepts change as they are explored and understood. For a time (i.e. when you are working within an accepted paradigm or conceptual system) it may appear that they have been pinned down, but very regularly in science and mathematics there will be a new definition that is proposed for a well-known concept. This new definition is the result of looking at the old situation in a new way. As a result, the concept has changed or it might be more accurate to say that our relationship with the concept has changed.

(5) The "law of non-contradiction" which states that it is impossible for **P** and **not P** to be simultaneously true. For example, an integer cannot be both even and odd since even means non-odd and vice versa. Note that this does not mean that either **P** or **not P** must be true, nor does it mean the stronger statement that one of the two must be provable.

Neither rule (4) nor rule (5) are as obvious as they seem. If reality is made up of processes and not objects, then the law of identity is not true but merely an artifice that we must assume in order to get started with any intellectual or scientific analysis of the world. On the other hand, we saw that Rothenberg claims that a crucial step in the creative process is denying the validity of rule (5). It is equally true that from the point of view of development and/or learning, some or all of these logical rules may well be broken.

6.9 A Conceptual System for Mathematics Based on Deep Thinking

However the role of logic in mathematics goes way beyond the logical rules that the working mathematician employs in the course of her teaching and research. Mathematics, when viewed as a vast conceptual system, has come to identify itself with its logical, deductive structure. Thus a mathematician will tell you that she "only proves theorems" when in fact, she does a great deal more than that. For example, she also is the creator of original concepts, structures, and ideas. Thus formal logic has a double role in mathematics. At what one might call the local level, it is a language and an organizational tool in the proofs and theories of mathematics. Globally it provides what many think of as *the* defining characteristic of the entire subject. Thus it is inevitable that my contention that there are important non-logical elements in mathematics will be controversial to some.

Ordinary logic has its value and its limitations. Its role is to tie together the elements of a theory or argument into one linear, coherent whole. This is wonderful when you are operating within a fixed paradigm or conceptual system. In fact, the main factor that makes something into a paradigm or conceptual system is precisely that it forms a coherent whole. It is this coherence that imparts stability and makes the system so resistant to change. It makes so much sense that one resists giving it up or even modifying it in a substantial way. Because mathematics uses logic in the two ways that I mentioned above, it is especially resistant to change and is the reason why what is in fact a conceptual system appears to many to be objectively "true."

This explains to some degree the power that mathematics has assumed in many aspects of modern culture. Therefore there is a general tendency to mistakenly identify mathematical models with reality (at least until the edifice comes crashing down such as in the recent financial collapse). Remember that a system doesn't so much describe reality as it creates reality, so that giving up such a system or even admitting its limitations entails, for many people, a frightening step into the unknown. It is well known, for example, that people with psychological difficulties are often reluctant to give up their dysfunctional ideas or behavior even when these clearly do them harm. As paradoxical as it sounds, such people are reluctant to let go of their problem because this is what seems most real to them—what they identify themselves with—and without it they would not know who they are. Their "problem" functions in the same way as a conceptual system; with it, things may be unpleasant but they "make sense." Ordinary logic is the glue that holds conceptual systems together in general, and in mathematics this is true to an extreme degree.

For many people mathematics *is* logic and logic *is* truth. It follows that the role of mathematics and quantification in our culture is a privileged one. The negative side to this is that the very coherence and compelling nature of mathematics that comes from its identification with formal logic makes the formalist conceptual system for mathematics especially rigid and resistant to change. This in turn imparts rigidity to those elements of modern culture that use mathematics as a model. This book is an attempt to break down this rigidity by highlighting the non-logical elements, which can be easily found within the edifice of mathematics but which the current conceptual system renders more or less invisible.

For every conceptual system there comes a time when the system is overwhelmed by a critical mass of new phenomena and problematic elements that do not fit within the old paradigm. In recent years, the classical view of what mathematics is has been challenged by techniques and results that mostly arise from the use of the computer. The moment will come, or perhaps has come already, when the coherence of the old view of mathematics will break down. Even then, many will hold on to the security of the old, but inevitably there will be people who are prepared to admit that the old system does not work and who cast about for something new. This book is an attempt to anticipate elements of a new conceptual system for mathematics. At the very least a new way of looking at mathematics will entail breaking the identification of mathematics with formalism and logic.

Any creative breakthrough always comes about when someone has the courage to break free from the system through which they establish what is real, but in the case of mathematics (and science to a lesser extent) such a break will be even more difficult than it is in other situations. This is because logical argumentation plays a stabilizing role for our entire culture.

The leap to a new view of mathematics will inevitably depend on reassessing the roles of logical and non-logical elements. We might begin with the simple realization that non-logical elements exist. They are, and have always been, a part of mathematics. For example, mathematical concepts always admit multiple representations and so are generally ambiguous. This kind of ambiguity is the motor behind the power and flexibility of mathematics. The constructive ambiguity present in situations of multiple representations is present in much of mathematics. For example, every tautological theorem, that is, every theorem of the form **P** if and only if **Q**, is ambiguous in the sense that it states that one element of mathematics can be represented in two different ways, **P** and **Q**.[viii] Thus ambiguity is an essential element in the theoretical structure of mathematics. Of course it is also present in the process of learning or creating mathematics where, as I have pointed out on numerous occasions, it is often the essence of what is going on.

Accepting the ubiquity of such non-logical elements in the local structure of mathematics should lead to a reassessment of the global nature of mathematics. The relative roles of formal logic and non-logical factors like deep thinking would then be reversed. Deep or creative thinking is how you think up or understand the ideas of mathematics; logical thinking is how you verify that your ideas are consistent and coherent. In this way, deep thinking would be primary and logical verification would be seen as secondary. Mathematics as a whole would then come to be seen as the creative endeavor that it undoubtedly is and not just as an exercise in pure logic. Just as Formalism makes logic into the defining characteristic of mathematics, deep thinking might come to be seen as the defining characteristic of a new and expanded conceptual system for mathematics.

The new is produced by a way of using the mind in a situation that cannot (yet) be made coherent, much less captured in a logical form. Today what the "new" mathematics will be cannot yet be conceived in any definitive way. From the point of view of formalism or the normal conceptual system through which we mathematicians explain to ourselves just what it is that we do, what I am trying to do may seem to be impossible. Mathematics for these people would not be mathematics if the logical requirements were to be loosened.

On the other hand this entire book has been describing the kinds of situations that produce deep thinking and this is the situation in which we find ourselves today in mathematics. Deep thinking is the essence of mathematics. It is a definite process that is characteristic of learning and creation but its rules are not the logical rules that we learned at school. However we cannot even get started with a consideration of deep thinking and its "process logic" without letting go, at least for a time, of the straightjacket of ordinary logic. In just the same way, we cannot conceive of what mathematics has become today without getting out of the straightjacket of considering formalism, formal logic, and the deductive system as the defining characteristics of mathematics.

Only by acknowledging the limitations of logical inference can one be receptive to more basic creative and developmental processes. Creative thinking arises out of the problematic and involves accepting the encounter with the darkness of "not knowing." It comes into being in a moment of discontinuity and not by creating a "logical" model of the situation that will enable one to replace the tension and opaqueness of the creative process with the clarity of a process that is algorithmic.

As the philosopher Hannah Arendt[ix] pointed out, the birth of a new idea does not come out of manipulating logical inferences. When one is "holding contradictory ideas simultaneously in the mind," one is in a kind of double bind—"this is the way it is but it cannot be like this." Facing this kind of dilemma, grappling with the impossible in this way, results in a mind that is active and aroused, a mind that is unable to settle on any of the possibilities that it is familiar with. On the other hand the mind that is spinning out formal inferences according to some set of rules is only minimally active. Logical manipulations, no matter how complicated or subtle, never challenge a person in the same way as confronting the need for a complete overhaul of one's system of beliefs.

Logic is the key element in the current conceptual system for mathematics, but mathematics contains much more than the current conceptual system allows us to see and comprehend. Just as logic functions today in a way that is both local and global, a new conceptual system for mathematics would use creative thinking in two ways. The first way would be the local usage, as the key element in the learning and creating of mathematics. Globally, deep creative thinking might come to be seen as a defining characteristic of the subject in the way that logic is today. In making this kind of fundamental transition, creative thinking would be used in yet a third way since it is the way the mind is used in moving from one conceptual system to its successor. The old system, formalism, identifies mathematics with formal logic. Its successor will be based on a new way of thinking about mathematics, centered on ideas, the creative process, and deep thinking.

6.10 Conclusion

If it is indeed true that there is no mathematics that is not cognitive, as Poincaré stated, and which seems pretty self-evident to me, then there are plenty of important implications that follow. All of these involve seeing through the smoke screen of absolute objectivity, certainty, and precision through which mathematics is traditionally viewed. These characteristics are not so much an intrinsic part of mathematics as they are of the current conceptual system for mathematics. They are a manifestation of the role that mathematics has come to play in our culture. Research papers in science are validated by statistical tests, politics is conducted via public opinion polls, and decisions in business and industry are arrived at via the creation and the application of mathematical models. When we look to establish the truth we tend to reach for mathematics.

But change and evolution are the rules of existence and mathematics is no exception to that rule. Mathematics is a creative endeavor that is grounded in psychological and social realities. This may seem to put limitations on mathematics and indeed it does break the identification of mathematics with truth. However, mathematics, looked at in another way, reveals an even deeper and more vital truth. Mathematics demonstrates the unity of mind and the natural world—the remarkable fact that the mathematical mind is able to see so deeply into the structure of the natural world but also that mathematics is able to reveal the properties of the mind.

---

[i] Poincaré (1909) quoted in Miller (1984).

[ii] The eminent mathematician and founder of Formalism, David Hilbert.

[iii] Consider the advocacy of experimental mathematics by people like Borwein (2008).

[iv] According to Euclid, "A straight line is a line which lies evenly with the points on itself." But one would have to know what a line is to understand this definition.

[v] Kanigel (1991).

[vi] Hersh (2014).

[vii] See also Grosholz (2007).

[viii] E.g., the theorem "a number is rational if and only if it can be represented by a finite or repeating decimal" places the rational number in two contexts, namely, the fractions and the real numbers.

[ix] Arrendt (1966).

Chapter 7

Mathematics as Process[i]

7.1 Introduction

What does the discussion about creative intelligence, deep thinking, and conceptual systems tell us about the nature of science? I shall continue looking into the nature of mathematics but the implications of the discussion will extend beyond mathematics. I have said earlier that we can't look at mathematics as merely the contents of a formal theory, from axioms to logically derived conclusions. Such a view does capture something important but it also misses a great deal that is vital.

7.2 What Mathematicians Do

Creative intelligence is a process. It is a process that produces mathematical concepts and mathematical objects. It is the process that is primary, that is, *mathematics is what mathematicians do!* Sounds simple enough but then again what is it that mathematicians do? Well, they do mathematics. So what does this circular reasoning do to advance our understanding of math? In mathematics circular thinking is frowned upon and it may indeed be a defect in the formal mathematical discourse that you might find in a textbook or research journal but, in trying to fathom what mathematics is all about, I suggest that it is not a problem at all. In fact, circularity may actually point us in the right direction. It may help us to see that mathematicians and mathematics, mathematical process and content, are fundamentally intertwined. They are two perspectives on one and the same phenomenon. It is a matter of convenience and perspective as to which one we choose to consider at any given time.

Remember that the great mathematician Bill Thurston said, "What we (mathematicians) are doing is finding ways for people to think about mathematics,"[ii] which is circular in much the same way as my statement above yet says something important about the interrelationship of mathematics and mathematical activity. After all mathematics is a creative process in which new results lead to new questions which lead to further progress in a kind of endless spiral.

I spent the good part of a book[iii] discussing the kind of situation in mathematics in which there are two equally valid perspectives on one situation where the two perspectives appear to clash with one another. In the present situation the two perspectives are process and structure. I referred to this kind of situation as ambiguous, an ordinary word that I appropriated to describe a fundamental aspect of mathematics that is normally overlooked.[iv] Normal ambiguity is viewed negatively and suppressed but I chose to emphasize its constructive aspects. Coincidentally a terrific book by Emily Gottshalk appeared in the same year as mine which was also about such positive ambiguities.[v]

In this chapter I intend to investigate the process/content ambiguity of mathematics; the ambiguity of the mathematician and the mathematics she creates. There is a considerable amount of confusion concerning this situation that I hope to shed some light on. Mathematics is not just about stuff; it is also about how the stuff comes about, how it is understood, and how it is learned. Process and content appear to inhabit logically distinct domains but the point of view taken in this chapter is that they cover the same ground and thus should be thought of as complementary—two valid ways of describing the same thing. Each perspective brings its own insights.

However these two points of view do not currently have the same status. For the most part, mathematics is considered to be identical to its content. Thus it is unconventional, to say the least, to consider the perspective that would take "process," the doing of mathematics, as primary or that we should consider process as being prior to content. Of course "prior" here does not necessarily mean chronologically so much as it refers to its ontological status.

Let us consider for a moment what it would mean to prioritizes the process dimension over the dimension of structure. It would root mathematics in the activity, the doing, or the learning. How would this work? For one thing it might force us to expand our definition of mathematics to include what might today be called informal mathematics as I discussed in the previous chapter. In addition to the usual formal world that consists of objects and structures, there would also be an informal world that forms the ground out of which the formal is created. The form, then, would be the crystallization of the informal. Objects would arise from process through what is called reification,[vi] for example, the process whereby infinite summation becomes a real number or subtraction becomes an integer.

Of course, this is not the way we learned to think about mathematics. It's the opposite of Platonism where you have ideal objects at the most fundamental level. In fact it's not the way we think about science nor, for that matter, the way we think about the human condition. We tend to think of objects as primary and some of these objects as elementary, the "atoms" of our theory. We are most comfortable with these sort of "atomic" ways of organizing the world.

The objects, whether they are actual atoms or else molecules, cells, people, or numbers are the primary focus. We start with elementary objects and then consider their relationships to each other and the ways in which they combine and interact to produce new, more complex objects. That is the way we structure the world. Even our language (but not necessarily every language) is structured in this way. Things (nouns) appear to come first and thus have priority over actions (verbs). We imagine that the world has this structure and that language reflects this reality but it is also possible that the reverse is true, that language comes first and the structure of language determines how we think and has given us a theoretical approach in which the world and various scientific disciplines are structured in this object-oriented and atomistic way. For the most part this way of looking at things has been handed down to us from the ancient Greeks.

7.3 Looking at Mathematics from the Perspective of Process

The alternative, a process approach, has something in common with the metaphor of Quantum Mechanics in which it is insufficient to think of an electron, say, only as a particle, that is, an isolated object with clear boundaries and a definite location in space-time. It takes the collapse of the wave function to "create" a particle, that is, to bring it into being. The wave is the particle before the collapse of the wave function.

There is also a "collapse" in the process of reification. Process, which is open and fluid, collapses, or crystallizes, into the reified object which appears to have definite boundaries or a formal definition and therefore seems to be solid and permanent. The process world is chaotic and that is how things are "at the beginning" in the informal dimension of the mathematical adventure. Process, which has no permanent form, certainly exists, but not quite in the same way as form does. The manner in which it exists resembles in some ways the process of informal thinking or free association where ideas keep flitting by and we have to be very alert to grab the significant ones and write them down so as to give them a kind of permanence. We sometimes do this by giving names to those ideas that we deem to be significant, and this highlights the connection between language and the process of giving structure. When most people use the word "mathematics" they normally are talking about the world of mathematical structure. When I use the word I will mean both worlds: the fluid one of process and the solid, quasi-static world of structure.

This fundamental assumption about how mathematics and, more generally, scientific disciplines should be organized is ultimately in question here. In line with modern Quantum Field Theory, for example, maybe objects are not the fundamental and permanent building blocks of reality. Objects change—even what we mean by specific mathematical concepts like continuity change; they come into being under certain circumstances and change or even disappear when they are no longer relevant. If change is fundamental and universal then one key question would be the manner in which objects, including mathematical objects, come into being for the individuals doing mathematics, for the larger culture of mathematics, and for society in general. Thus the manner in which a process becomes an object, reification, is central. However there is also the inverse process, namely, the manner in which structure disintegrates, is modified, or gives rise to process. That process can come from structure is simple. When you have numbers you can then do arithmetic. But how do structures change? How does a simple structure give rise to a more complex one? One way this happens is in response to problematic elements that inevitably crop up in any theory and only appear after the structure is investigated by the mathematician. Thus, you cannot divide in the system of integers. Or else there are the famous unsolvable problems in Euclidean geometry such as the trisection of the angle. These problems reveal certain inadequacies of the system in question. Their resolution often requires the creation of a new system that extends the old. But before the new structure can appear it is necessary that the way of thinking that comes with the old system, and in which the theorems are considered to be absolute truths, breaks down. Only this provides a space in which a new way of thinking can emerge. In this new theory the original truths are seen as relative (to the set of axioms used, for example) and not absolute. A good example of this is the movement from Euclidean to non-Euclidean geometry or simply making ratios into rational numbers.

Thus there is reification, which is the way structure is constructed and what you might call "deconstruction," the melting down of old structures to make room for new ways of understanding. One goes from process to content, the other moves in the opposite direction, from structure to process, and then to new structures. Both directions involve mathematical creativity, although deconstruction and reframing are not always recognized as fundamental aspects of mathematics. It is important to bear in mind that both construction and deconstruction are human activities. Reification involves the birth of concepts, deconstruction is essential to reframing an old concept in a new way.

Let us think for a moment about what it would mean to take "mathematics as process" as a place to begin the discussion of the nature of mathematics. Of course, this would not mean that mathematics is only process. It means that you must start with process (and with an understanding of reason which I go into below) when you describe mathematics. You don't start with mathematical objects such as number systems, for example. You don't even start with axioms, not even with a system of logic such as the propositional calculus. These traditional approaches view process as, for example, the unraveling of the properties of mathematical objects that are implicit in an axiomatic system. Learning and creativity are bracketed off from "real mathematics" and only considered in separate disciplines such as education and psychology. Not even the history of mathematics gets treated as a "mathematical" discipline and so even advanced mathematics students tend to be quite ignorant of the history of their own discipline. We tend to ignore process in our teaching so students get little training in problem solving or even training in the process whereby one becomes fluent with a difficult mathematical concept. As a direct consequence of the mathematical community's approach to the nature of mathematics, process is neglected and even reason is not seriously investigated other than to identify it with logic. There is a sense in which all traditional approaches are fundamentally static.

Yet at its very origin, mathematics is an activity: counting, measuring, comparing, estimating, and so on. For example, we might be interested in investigating "quantity" or "amount." We begin by trying to answer the question, "Which is more?", or equivalently, "Which is larger?" To consider such questions, it is necessary to structure quantity by inventing a language. We require abstract carriers of quantity and that is how we get to the concept of "number." "Number" is a reification of the various activities that were mentioned above.

Even though it may not be possible to go back in the history of mathematics to isolate the first emergence of number, the development of the number concept in infants will give us some hints. As we saw in Chapter 1 the situation is surprisingly complex. It seems that infants (and some animals) have a latent numerosity, which could be defined as a rudimentary ability to recognize number, a built-in capacity to develop the number concept. This does not mean that this latent capacity need necessarily be activated or developed in the same way that other genetic traits can remain latent until some environmental stimulus turns them on, so to speak. So the capacity for number leads to various activities that require an enriched and refined sense of number, which leads to a more sophisticated number sense in a spiral of development that ends up with the very complex number systems with which we are familiar.

It is interesting that educators have long observed that students learn by doing. Then they learn to give names to bits of mathematical activity, creating in this way a structure which becomes the scene of further, higher level, activity. Thus a fraction starts as a problem in long division and ends up being a number just as the negative numbers arise from the process of subtraction. Thus processes get reified out as conceptual objects. Then in the formal recapitulation we start with the objects and reintroduce processes of various sorts. We are reluctant to begin with process because objects seem so much more precise and solid. Of course a given process could conceivably be reified in more than one way and lead to different formal systems. Possibly we move in certain directions because of certain tendencies that are built-in.

As a quick aside here, reflecting on quantity, the idea of more or less, leads directly to the idea of "equality," that is, neither more nor less. There is no place here to go into the idea here but let's just flag it as a key idea in mathematics, science, and even language.

There are other ideas that flow directly from ordinary activates that find their way into mathematics. Ideas such as "chance," "continuity," "nothing," and "infinity." All of these arise out of activities, out of solving problems. These and a host of other questions begin at a level that is concrete and experiential; they arise in the ordinary activities of human beings.

Thinking about mathematics in this way may make some people a little uncomfortable, maybe even bring on a feeling of disorientation. We have become habituated to the metaphor of mathematics as an enormous and intricate edifice of eternal truth. In this view even though mathematics may have been created by human beings, nevertheless it transcends the human condition and inhabits a mystical, Platonic realm that lies beyond time and space. We love the idea that even if individual human life is finite, the mathematics that we create will live forever because its truths are eternal and immutable. It's a comforting place to inhabit whereas mathematics as process brings us back to the contingency and uncertainty of life, something that may appear to recede as we contemplate the grand structures of the great mathematical theories.

7.4 Mathematics is What People Do

To go back to the statement "mathematics is what mathematicians do," note that it has an emphasis on process but also emphasizes that mathematics is an activity that is carried on by human beings. The late neurologist Oliver Sacks said,[vii] "Science is a human enterprise through and through, an organic, evolving human growth, …" The same could and should be said about mathematics.

It is particularly appropriate to look at mathematics from this point of view. The article from which this chapter is taken appeared originally in a volume that celebrates precisely the movement from an emphasis on structure to an emphasis on activity. This change can be traced back to two remarkable books: *Proofs and Refutations*[viii] by Imre Lakatos and *The Mathematical Experience*,[ix] by Philip Davis and Reuben Hersh. Both of these books turned our attention away from the structure and content of mathematics and towards what is actually going on when mathematicians do mathematics. This shift involves moving from a view of mathematics as consisting of static, permanent structures in the form of definitions, theorems, and proofs, as well as algorithms and various methods of computation, to mathematics as a dynamic, creative, and thus human process. This necessitated a critique of the prevailing framework for mathematical activity, namely formalism and Platonism, and a search for its replacement of which this chapter is an example. In short, it is an attempt to describe mathematics as a theatre of creative intelligence.

7.5 Understanding and Learning

Thurston and many others have identified "understanding" as *the* basic mathematical activity. Mathematicians try to understand the mathematical phenomena that they encounter in their professional lives. But what do we mean by understanding? I recall a comment made by a student concerning some very dry derivation of the consequences of the axioms of a topological space, "I follow the argument but I don't understand it." Following involves a step-by-step process of verifying that each step follows logically from the previous one. Understanding is more global or holistic. It usually happens suddenly and may happen more than once for the same material, that is, there are depths of understanding. When it happens one might say something like, "Aha! I see what is going on here." Understanding a proof, for example, often means that the idea behind the proof is revealed. Even a fairly complex proof may be the elaboration of an idea which can be stated quite succinctly. But understanding a subtle mathematical idea like real numbers, continuity, or randomness is a significant accomplishment that, in a certain sense, is never complete. There is a continuum of understanding which moves from the first glimpses of understanding by the young student to the sophisticated understanding of the research mathematician.

It is worth emphasizing that both research and learning involve understanding. Doing mathematics involves learning (about) mathematics. We would have a much more profound appreciation for education if we realized that learning at the most profound level involves the same mechanisms as doing research. (Of course not the same level of complexity.) Learning is not the passive accumulation of facts. Learning involves mastering ideas, that is, conceptual understanding. Learning and understanding are both creative activities; they are a manifestation of creative intelligence.

Though mathematicians and math educators do not normally communicate with one another because they live in different disciplinary compartments, educators are well placed to understand the difference between being content or process oriented. Educators and teachers need to focus on process because it is evident that learning is a process. Today, influenced by a view of knowledge as mere data, there is a tendency to think of learning as the accumulation of facts and techniques. But this is not what we mean by learning or doing mathematics and has little to do with becoming fluent with a particular mathematical world.

Remember the first-year graduate student in one of my classes who, when asked what mathematics he had studied, answered that he had read such-and-such a book on analysis and could quote the theorems and proofs from memory. Very impressive! Unfortunately he was unable to answer or even approach the questions in the first assignment because it turned out that he had never actually done a problem before. This student knew no mathematics at all and this illustrates what I mean by saying that mathematics is not primarily about content. Both learning and research are processes whereby human beings "find ways … to think about mathematics," where "thinking about" could be taken to mean "understanding."

The shift from viewing mathematics as structure to viewing it as process is a radical change in perspective. We were all brought up to consider the content of mathematics as what is real and defines the subject. It is as though every area in mathematics could potentially be written down in some sort of definitive textbook which, of course, would never be complete. In this view, a subject would consist of a compendium of facts, and research involves adding to this collection. In this view, learning would involve getting familiar with those facts in the manner of the graduate student I caricatured. Of course this is not what learning means at all. It really means becoming fluent with the material, understanding the main ideas and concepts, and being able to apply them in an intelligent way. Just as fluency with language implies the ability to produce sentences that may have never existed before, fluency in mathematics means being able to work with situations that one has never encountered previously.

This shift from facts to processes is crucial for the way we think about mathematics and science (and the world in general) because facts seem to be solid, objective, and potentially eternal. Process, on the other hand, carries the connotation of impermanence, of coming into being and then disappearing. Process also seems to have elements of subjectivity. There is always someone who is creating or learning. Of course we could imagine that the agent behind the process is an algorithm or computer program which will lead us to consider the possible role of artificial intelligence in the process of mathematics. I shall have more to say about this later on in the book.

7.6 The Process of Reason

Of course, mathematics is not just any process. It is a rational process and, in particular, it is an intelligent process. People *do* mathematics; they *learn* mathematics. So in order to find a way to talk about mathematics as process, we must define what we mean by a rational process. What is reason? I'll give a kind of definition a little later on, but whatever reason turns out to be, it must involve intelligence, that is, mind. And, as far as we know, this is why mathematics is a human activity—mathematics is about *people* learning to think about mathematics.

It follows that what I have called "the dream of reason,"[x] (cf. *How Mathematicians Think*, p. 12 or 84) does not hold, partially, as we shall see, because we misunderstand the word reason. We identify reason with logic and divorce it from intelligence. Then we imagine that "reasoning" involves following the rules of logic, in particular, that it is rule-based.

The traditional dream of reason is, of course, that we can produce mathematical and scientific theories that are certain, objective, and identical to what they model. In particular we imagine that a theory is static and eternally true. We feel that truths can never and will never change. In the dream of reason, the theory stands alone liberated from reason, mind, and humanity. Understood in this way, the dream of reason is a bad dream because the humanity of the dreamer has been lost or ignored. If this dream is not valid, there is no "final theory" of physics, economics, or any other science; there is not even a theory of the stock market that works over an extended period of time. In our subject it is clear that mathematical activity will continue without end.

This chapter will propose a new way of thinking about rationality as process. On this basis we might be able to rehabilitate the dream of reason and reclaim it for humanity, instead of as a way of replacing human beings with machines or sophisticated artificial processes.

7.7 Formalism as Process

As a precursor to a discussion of the rational process, let us consider the nature of formalism. Formalism identifies mathematics with a process, namely, the axiomatic method that operates within the context of a set of axioms, definitions, and operational rules of logic. It is still the dominant way of thinking about how doing mathematics works, and so let us go back and revisit the nature of an axiomatic system and the process of working within it.

Here are a few questions regarding axiomatic systems that reveal the limitations of the way we normally think of it.

1. What does Thurston's comment mean when applied to an axiomatic system, that is, what does it mean to understand such a system as a whole or even to understand some particular theorem that follows from the axioms? In particular where is understanding to be found—is it part of the system or extraneous to the system? In other words, what is understanding and where is it located?
2. Given a choice of some particular set of axioms, why are some results interesting while others are not? Why do we often make statements like, "That result is deep," or "This result is trivial?" Can the basis for the judgments "deep," "trivial," and "interesting" be found within the system or are they outside the system? Or are they merely subjective with no objective meaning? Are these judgments even part of mathematics at all? If they are not, then how do you decide whether or not to publish some article in a prestigious research journal?
3. Suppose you are trying to prove a result which has been found to be plausible empirically. Where does the idea for the proof come from? Does the mathematical idea exist within the system or is it extraneous to the system? In general, what is the relationship between the concepts and ideas of the subject and the axiomatic structure?

These questions and others demonstrate that when we discuss the axiomatic system, we focus on certain aspects to the exclusion of others. There is a great deal more going on than mere deduction from axioms. Another way of putting this difference consists in saying that the rational process involves much more than applying logical inference to a set of *a priori* truths.

What I have to say will be clarified if we take a new look at what is happening in the axiomatic system. In order to make things concrete I shall take a quick look at Euclidean geometry as a prototype of an axiomatic system and give a more complete account of what is going on. What are the answers to the three questions I listed above in the case of Euclidean geometry? What, in short, does it mean to look at Euclidean geometry as a rational process?

7.8 The Discovery of Reason by Doing Euclidean Geometry

Euclidean geometry is one of the foundational elements of Western civilization and the root of the technological revolution that is currently sweeping over us. I began rethinking Euclidean geometry when I was teaching the subject to my thirteen year old grandson during the COVID lockdown. Euclidean geometry is not often taught in schools any more but friends and colleagues who are scientists and mathematicians all concurred that the subject had been important to them in their intellectual development. Why was that? Why study Euclidean geometry?

Most people regard Euclidean geometry as the prototype of a deductive system—definitions, axioms, and theorems deduced through pristine logical thought. They may believe that the subject is algorithmic so that it could be done by a computer. In other words they imagine that Euclidean geometry could be done without human intervention. So is Euclidean geometry a matter of pure deductive logic?

The significance of Euclidean geometry to the mathematician goes beyond its theorems. It includes the means through which these results are obtained. Euclidean geometry was such a significant part of the education of scientists, physicians, and mathematicians of my generation because it taught us how to think. In particular it showed us that thinking involves a deep connection between logic, intuition, and insight.

Even though Euclidean geometry is held up as the prototype of a deductive system, which may appear to be abstract, in reality it is concrete and geometric. It is concerned with the structure of two- and three-dimensional (flat) space. In fact for the Greeks, Euclidean geometry *defines* space. It is not only about proving theorems, but also about developing an intuition for the geometry of space and fitting these intuitions into a system of verifiable truths. Doing Euclidean geometry involves logical thinking and it also involves the nonlogical procedures of intuition and insight that I will describe below.

Imagine that you are studying Euclidean geometry from scratch. Imagine that it is an area of research. "Doing" Euclidean geometry would work like this: First of all you have to think up some geometrical statement (or potential theorem) which is interesting and accessible on the basis of current knowledge (the axioms you have agreed on and the theorems you have established so far). This is akin to deciding which hypothesis to test in a scientific experiment. Then one has to guess whether the statement is true or not. If you guess "no," you try to find a counterexample. If you guess "yes," you try to construct a proof. The proof itself does not come into existence through a series of purely logical statements. A proof (or counterexample) is built around some idea (which may turn out to be a geometrical construction). In other words, you have to know *why* it is true before you try to prove it. Actually the idea for the proof is usually what you mean by "understanding" the result.

Where does the idea or the understanding come from? Obviously it comes from the mathematician. It is only at this stage, after the idea has revealed itself to you, that you attempt to write down a proof and this is the only step in the procedure that is strictly logical.

Thus doing geometry involves **intuition** (developing the hypothesis and guessing true or false), **insight or creativity** (coming up with the hypothesis and the idea for the proof), and **logical argumentation** (for the purposes of verification and communication).

All three are essential parts of a deductive system and this is how it feels to work on such system *from the inside*. This is how it feels to *do* mathematics. The further you go the less value you ascribe to the third, logical, part. Thurston[xi] said, "When the idea (behind a result) is clear, the formal setup is usually unnecessary and redundant." Creative scientists are basically interested in new insights, original ideas.

7.9 What is Rationality?

Let us *define* the rational process by abstracting my description of Euclidean geometry. Thus, *rationality involves intuition, insight, and logic.* Logic is just one of the steps not the whole ball of wax. This gives us a working definition of the rational process. Intelligence and especially creative intelligence is very subtle. It is the subject and theme of this book. Rationality as I've been describing it is one way in which creative intelligence manifests itself and it often involves deep thinking. When we struggle to get a grip on creative intelligence let us think about the rational process that I have been describing. If creativity is the very essence of science and mathematics, then it is the heart and soul of the rational process. Deep thinking is the rational process in action.

7.10 Intuition or Fast Thinking

Intuition is thus an essential aspect of the process of reason. The Nobel prize winning psychologist and economic theorist, Daniel Kahneman, is famous for demonstrating that the economic actor is not only logical.[xii] In fact according to Kahneman human beings are capable of two kinds of thinking that he calls fast and slow. Slow thinking is what many people think of as logical thinking. They forget that fast thinking, which is involved in intuition, is also essential to rational thought. Fast thinking refers to what many people call their "gut," and there are many disastrous example of the damage that can occur when fast thinking is unchecked by slow. You can read the history of rational thought as the attempt to control fast thinking with slow thinking but, in my opinion, that would be a mistake, for reason has room for both of these modes. Perhaps we should follow the example of behavioral economics by admitting that rational thinking involves a synthesis of intuition and systematic, rule-based procedures.

7.11 Reason or Revelation?

I have heard it said that the Ancient Greeks contributed their ideals of reason and harmony to modern civilization. The Ancient Hebrews, on the other hand, contributed the ideal of revelation, which you can think of as the voice of God speaking to Moses from the burning bush, but more prosaically, as the discontinuous emergence of a new and powerful idea. In other words, the aha! experience. Anyone who has done any research in mathematics or even mastered some nontrivial mathematical concept like real numbers or limits will have had this moment when things begin to come together and make sense.

The confusion I mentioned above concerning the true nature of the rational process comes from reducing reason to logical inference, on the one hand, and ignoring the role of revelation on the other. Logic is rule-based, orderly, and predictable, whereas creativity is not predictable and does not seem to follow any system of *a priori* rules. The problem with reducing reason to logic is that what I (a bit dramatically) called revelation is an essential part of doing mathematics and even learning mathematics. Revelation is an essential part of what I've been calling "deep thinking" in the earlier chapters and which we can now see is an essential aspect of reason. The way towards a more compete view of mathematics involves accepting that reason is a lot more than logic and includes both intuition and insight. Only then will it be possible to respond to Poincaré's question—the perpetual mystery of the power of mathematics to penetrate so deeply into the natural world. This power comes from mathematics' ability to harness all aspects of the rational process.

7.12 Creativity and Insight

Creativity is not to be confused with the production of what is new. It involves insight, by which I mean the discovery of a new way to "see" some situation. We understand some event or situation by placing in a context, that is, framing it. Then the most basic creative act involves *reframing*, that is, coming up with a new way to understand a given situation or event. So "getting the idea" often means finding a new point of view from which to look at some situation. In science a frame is called a paradigm and reframing is referred to as a paradigm change. I am thinking, for example, of the way Einstein reframed our understanding of gravity.

But to remove the idea of creative reframing from the rarified atmosphere of genius, let me refer to the example discussed in Chapter 1 that applies to everyone. As discussed there, an early conceptual task for children consists of combining the two systems elementary number subsystems that they are born with into their first learned number system—the system of the counting numbers: 1, 2, 3, …. That development is a child's first creative learning experience in mathematics. However, because we are here concerned with the rational, I want to focus on the next step children take a few years later—the reframing of "number" from the frame of the counting numbers to the new frame of fractions or rational numbers. Recall that this involved moving from thinking of two-thirds, for example, as a ratio such as when a pie is divided into three parts and you choose two of them, to thinking of two-thirds as a number in its own right on a par with the positive integers.

7.13 Creativity as Reification

I already discussed the crucial but non-obvious question of why 2/3 is a number at all. If it is a number then by what process does it become one? The child must come to see 2/3 as a single object, that is, she must reify the ratio of two to three into a new kind of number. When this happens with respect to two-thirds, the child can do the same thing for other fractions. She has then undergone a total conceptual reorganization, a reframing of her understanding of number. Reification of pairs of whole numbers, which brings the fractions into existence, is nothing less than a paradigm shift and yet almost every child goes through this shift sooner or later. Of course we don't call it reframing, we call it learning. We have all been there but we have forgotten, and so for us it is "obvious" that fractions are numbers. We forget what a significant development this was. But there is no reason *a priori* that fractions should be numbers that extend the system of counting numbers. Moreover there is no reason why the set of ratios should make up a new number system for which the old operations of arithmetic still make sense. You can see what a big deal it is when you think about the extravagant claims that the Greeks made for the rational number system (which they seem to have identified with reason itself) and how profoundly shocked they were when confronted by the existence of irrationals. The creation of the rational number system in human culture or in the mind of a child is a prototype of an act of creativity.

One crucial point about this example. The two number systems, counting numbers and factions, are incompatible (or incommensurate, as the philosopher of science Thomas Kuhn would have it) with one another in the following sense. If you ask a child who lives in the world of counting numbers how many numbers there are between 2 and 3, she will say none. But a child who made the creative leap to the world of rational numbers will say that there are lots of them. And of course they would both be correct within their respective frames or conceptual systems. In other words correct and incorrect are *relative* terms, relative to context, of course.

Within a conceptual system like the integers or the rational numbers, one insists on consistency and thus the absence of things like contradiction and ambiguity. A well-designed theory has no place for such logically problematic things. However it is precisely these problematic elements that drive paradigm change, the movement from one conceptual system to another. In Greek geometry, there were "unsolvable problems" like trisecting an angle and squaring a circle. To solve them, you need to change the context of the discussion, that is, reframe the problem. An even more subtle problem concerned the status of the parallel postulate in Euclidean geometry. Questioning its status as an axiom and therefore "obviously true" leads ultimately to the development of non-Euclidean geometry.

One doesn't leave one's present paradigm willingly or easily; you need to be driven out by a problem that you cannot solve in the old system. For the move from the counting numbers to the rationals, the problem might be that the division of whole numbers is not closed within the system of counting numbers. In other words though you can divide 7 by 3 in the counting numbers (and get an answer of 2 with remainder 1), the answer is not a number in the system you started with.

This leads me to say that *creativity transcends logic.* It involves the sudden leap to a higher point of view. A problem that was intractable in the original frame becomes transparent when looked at in the new. It follows that rationality and deep thinking are not (only) logical thinking.

One further comment is brought out by this example. In mathematics there are many kinds of numbers: counting numbers, rational numbers, real numbers, complex numbers, and many others. This whole hierarchy is built on the primitive idea of "number." Yet "number" in the abstract is never defined in mathematics. We all have a feeling for number because (as we have seen) even a six-month-old child has two separate conceptual systems for number. Nevertheless this "feeling for number," what I called numerosity earlier, is informal and thus never defined explicitly. The specific kinds of numbers I mentioned are, in comparison, well-defined. In this way the conceptual world of mathematics (and physics) emerges out of an informal world. Number is not special in that regards. What I said also applies to the concepts of time, space, energy, randomness, and many other concepts. In fact all of the building blocks of science have informal roots. Mathematics lives in both formal and informal worlds. Intuition functions very well in the informal world and creative reframing often has to go back there, but logical processes live exclusively in the formal world where things have explicit meanings.

7.14 Strong and Weak Objectivity

The power of mathematics and science lies in their claim to "objective" truth. In this section I would like to take a few paragraphs to discuss the nature of objectivity. Most people believe in a kind of objectivity which I shall call "strong objectivity" and define below, which is analogous to what is meant by "strong AI."

We sometimes say that some phenomenon is "merely" subjective when we mean that it comes from personal prejudice or idiosyncratic opinion. Thus we would object to a mathematical theorem or scientific experiment being influenced by religion, race, or gender. Objective means that such matters, as well as many other cultural factors, do not influence the result. Of course some people consider mathematics itself to be a culture, but within mathematics the criterion of independence from arbitrary opinion might serve as a minimal way to differentiate subjectivity and objectivity—we could call this *weak objectivity*.

However there is another possible meaning of objectivity. Something is objective in this sense if it does not depend on mind or intelligence. You could say that it is objective in this sense if it would continue to be true even if there were no human beings around. This must have been the idea behind putting a diagram of the Pythagorean theorem on Voyager 1 in the hope that the truths of Euclidean geometry were universal and would be recognized by any intelligent being. Is the Newtonian theory of gravity objectively true? Most people would say that it is, even if that truth is only an approximate one. Is the relativistic theory of gravity objectively true? Is it an eternal truth? Maybe it is but the jury is still out. At any rate a scientific theory is objective in this sense if the theory exists independent of the scientist who formulates or studies it. So some believe that the scientist *discovers* what is already there, that the rules of the universe and of mathematics are built-in, so to speak. Let us call this *strong objectivity*.

I believe that mathematics is objective in the weak but not the strong sense. This is what I take away from the writing of Reuben Hersh in books like (1980) *What is Mathematics Really?* (Oxford University Press) and it is, I believe, the essence of the point of view that is celebrated in the volume where this essay originally appeared.[xiii] This implies that the truths of mathematics are human truths not Platonic truths. So the truths of Euclidean geometry are objective in the weak but not the strong sense. In exactly the same way one could subscribe to the hypothesis of "weak AI" and not "strong AI." Notice that the "weak" position depends on an essential connection between human thought and science; the "strong" position on the other hand holds that once the theory (or technology) is established, human beings are redundant. One holds that the process of rationality can operate independent of human beings, the other that human beings are the essential measure of rationality, for rationality is the process by which human beings understand the world and themselves.

7.15 Contradiction and Ambiguity

Because we conflate reason with logic and are oblivious to the role of paradigm change in mathematics, we tend to be hostile to the presence of contradiction, paradox, and ambiguity within mathematics, much less be open to their constructive uses. Most people believe that mathematics and science are free of ambiguity, contradiction, and paradox and yet I wrote a long and well-received book about the constructive uses of these things. To give an elementary example, the number zero was unacceptable to the Greeks because of its contradictory nature—something that represents nothing, as one author put it. The contradictory nature of zero remains but this has not stopped us from using it productively in science and technology. The same thing could be said for *infinity* (remember that mathematics has been called the science of infinity), randomness, and many other basic concepts of math, not to speak of the idea of complementarity in Quantum Mechanics which makes what appears to be a paradoxical situation into a fundamental principle.

The human mind has discovered creative ways to make use of contradictory elements. As I mentioned in the chapter on creativity, the legendary physicist Niels Bohr distinguished between two kinds of truth. "An ordinary truth is one whose opposite is a falsehood. A profound truth is one whose opposite is also a profound truth." Not only is he saying that something can be contradictory and true at the same time, but he is further asserting that an inner contradiction may be an indication of its profundity.

Thus there is a huge difference between what I have been calling human rationality and logical, rule-based argumentation. The human mind is extraordinary! It can create something profound out of nothingness or out of chance. It can make constructive sense out of what appears to be meaningless. It can create regularity out of the very lack of regularity, that is, randomness.

This gaping hole in the prevailing philosophy of mathematics arises because of the view that identifies mathematics with its formal structure and ignores informal mathematics. In order to accomplish this, it has been necessary to elevate rule-based logical argumentation to a kind of foundational principle, which is, of course, what formalism does. In this way, process, creativity, and learning, not to say history, have been banished from the way the subject is described and understood.

A key element in all of this is a misunderstanding of the meaning and the role of ambiguity. The conductor and composer Leonard Bernstein once said that the richness of music is proportional to its ambiguity, and the same can be said of mathematics. A measure of how deep a result is can be given by the number of ways it can be understood, the number of different contexts that it fits into. We continually look at particular mathematical concepts from multiple points of view. For example, a matrix can be thought of as an object, a rectangular array of numbers, or a process, a linear transformation. Notice that a reified structure is ambiguous since it is both process and object. And you choose to look at it from one or the other perspective depending on your need at any given time. In fact we think of the matrix in both ways at the same time and move freely from one context to the other. The point here is that the ambiguity, thought of as the ability to look at one situation from a multidimensional point of view, contributes to the power and creative potential of an idea. When you start to think of things in this way, one quickly sees that ambiguity crops up everywhere in mathematics, and moreover that it is a generator of understanding. Logic may have no place for ambiguity but rational thought and deep thinking use it all the time.

7.16 Logic, Algorithms, and the Human Mind

Rationality is best understood as the cognitive function of the human mind. One isolates and abstracts some particular aspect of cognition, logic for example, and then the case is made that one can reconstruct all of human cognition from that one property. There was the famous attempt by Russell and Whitehead to reconstruct all of mathematics starting with the rules of logic. Of course that attempt failed but it was interesting and was used for later developments like the Incompleteness Theorem of Gödel. In the long run, extreme reductionism never works. You cannot capture the full potential of the human mind using one particular model of mind. The mind is just too large and free for that to be possible. Logic and algorithms are one way the mind functions but not the only way.

The way to see this conclusively is to look at the history of science and mathematics. This history consists of the development of paradigms, conceptual systems, or ways of understanding aspects of the world. Inevitably, as the subject progresses, anomalies or inconsistencies develop. The response of the scientist is ultimately to develop a new theory that contains the old theory but expands it so as to explain the anomaly. Thus the non-rationality of the root of two is resolved by moving from the rational numbers to the geometric numbers, the problem of trisecting the angle or squaring the circle ultimately leads to the real number system.

The process of (human) rationality has room for, and even thrives upon, the problematic in the form of ambiguities, contradictions, and even paradoxes. It would be a great shame and a loss to our collective humanity to reduce human rationality to a process which eliminates the problematic and focuses exclusively on one particular form of thinking.

7.17 Conclusion

This chapter is my attempt to breathe new life into the question, "What is the true nature of mathematics?" When I think of mathematics, I think of a medieval cathedral like Chartres or Notre Dame. Magnificent! Fractal-like in its intricate complexity. But the magnificence of those cathedrals is the creation of thousands of anonymous and highly-skilled artisans. Looking at it at a deeper level they are the product of the creativity of the human mind. They are all examples of what I have been calling creative intelligence. So it is with mathematics. One cannot fail to be impressed, even amazed, by the beauty of mathematics and its ability to see so deeply into the structure of the natural world. When you look at the surface structure you are struck by its beauty and intricacy. But it is not static like a cathedral. It is dynamic, ever-changing, and developing. And that dynamism is the very essence of what is going on. When you come into contact with mathematics you touch the essence of creative intelligence! It is creative intelligence which is always dynamic. In the same way as a flower is the product of creative intelligence, and each human being is a manifestation of creative intelligence.

This is the reason I have been emphasizing the essential role of human activity in mathematics and elsewhere. Doing mathematics and learning mathematics both involve the same basic mechanisms. This capacity that we have as human beings for reason does not set us apart from the rest of the universe. Our reasoning ability is the very expression of the ocean of creative intelligence that surrounds us and of which we are a profound expression.

When we become aware of the sun, which I am using as a metaphor for creative intelligence, we see it everywhere. It is there in the great symphonies and the great works of fine art and sculpture. It is powerfully there in mathematics and the great theories of science. But don't be fooled by the magnificence of the cathedral into thinking of some sort of static, final theory. The essence is its dynamism! Creative intelligence is a dynamic process and so is mathematics. The deep nature of mathematics is process, which is creative intelligence in action.

---

[i] The ideas in this chapter are based on the paper *What Mathematicians Do: Mathematics as Process and Creative Rationality*.

[ii] William Thurston, *On Proof and Progress in Mathematics*, Bulletin of the American Mathematical Society 30, no. 2 (1994): 161–177.

[iii] William Byers, *How Mathematicians Think* (Princeton University Press, 2007).

[iv] It could also be called complementarity in the sense that the wave and the particle description of a subatomic particle are complementary in Quantum Mechanics.

[v] Emily Gottshalk in the book *Representation and Productive Ambiguity in Mathematics and the Sciences* (Oxford University Press, New York, 2007).

[vi] Cf. Anna Svard, *Reification as the Birth of Metaphor*, For the Learning of Mathematics 14, no. 1 (1994).

[vii] Quoted in the review of his posthumous book, *Everything in Its Place*, in the New York Review of Books, Vol. LXVI, No. 10, June 6, 2019.

[viii] Imre Lakatos, *Proofs and Refutations: The Logic of Mathematical Discovery* (Cambridge University Press, New York, 1976).

[ix] Philip J. Davis and Reuben Hersh, *The Mathematical Experience* (Birkhäuser, Boston, 1981).

[x] William Byers, *How Mathematicians Think* (Princeton University Press, 2007).

[xi] William Thurston, *On Proof and Progress in Mathematics*, Bulletin of the American Mathematical Society 30, no. 2 (1994): 161–177.

[xii] Daniel Kahneman, *Thinking, Fast and Slow* (Farrar, Straus and Giroux, 2013).
[xiii] B. Sriraman (ed.), *Handbook of the History and Philosophy of Mathematical Practice* (Springer Nature, 2024).

Chapter 8

Learning as an Act of Creativity

8.1 Introduction

In Chapter 5, I discussed the nature of creativity and then, in Chapters 6 and 7, how we might begin to think of mathematics as the manifestation of what I called creative intelligence, and as a result is best thought of as a process. I remarked that mathematicians are primarily concerned with understanding the phenomenon that they are studying. You could say that mathematical research involves learning about mathematics. In this chapter we will see that when we speak of the "learning" of the mathematician or the "learning" of students studying mathematics, we are speaking of what is essentially the same phenomenon. That is, learning is learning whether it occurs in the mathematician or the student. At the most important level, all learning is a creative activity and thus inevitably is accompanied by deep thinking.

Modern technology is making profound changes in the way people think and, as a consequence, in the way people learn. (And, of course, we are all aware of the changes which are right around the corner.) The nature of thinking and learning are undergoing basic and rapid change with implications that are extremely far-reaching. Many people, especially young people, are using their minds in a way that differs qualitatively from how the mind was used in the past. In other words the process of thinking and reasoning, and with it human consciousness itself, is changing rapidly and dramatically.

This change affects everyone but is most clearly evident in every home where there are children and in any classroom or university lecture hall. The environment in which learning happens has been changing for some time now but social and educational institutions have had trouble keeping up. Because technological change arises so rapidly society does not have the time to focus on the implications of these changes for education. How should the process of education be changed to respond to an atmosphere of continuous and accelerating change? Will formal educational institutions continue to exist, and if so, what should their role be in this new world that is coming into being? What are the appropriate goals of education? What does learning mean in today's society?

Learning cannot be separated from the different ways of thinking that were discussed earlier in the book. To each form of thinking there corresponds a kind of learning—systematic, logical thinking is associated with rule-based learning, the learning of facts and algorithmic procedures. Deep thinking is associated with a different form of learning, which I shall call *deep learning*. Deep learning is the most important kind of learning. It should be the goal of education at all levels but our educational institutions are, for the most part, either unaware of its existence or else choose to ignore it. This chapter will address certain basic questions about deep learning: What is it? Can it be taught? Should it be taught? What is its role in the world of information technology?

8.2 Learning

As I said above, learning is, in a sense, the factor that all of the processes that I discussed in previous chapters have in common. Learning is an extremely general and basic activity. In a way it is *the* quintessential human activity—the means through which civilization is developed and sustained and individuals are trained to play their assigned roles in society. However learning is not restricted to human beings. Human learning springs from the same sources as the learning that occurs in other animals and learning may well be even more general than that. I derived the notion of deep learning from work in human conceptual development but perhaps all development, the entire evolutionary process, for example, is a kind of learning, a kind of deep learning. If this is so, and I believe it is, then the process of learning is a central aspect of all life and all development. Learning is the way the world works and it is the natural way that we human beings work. It is the basic mode of functioning of mind.

Children have always spent much of their time learning but today even adults are forced to spend a great deal of time and energy learning and adapting to a shifting technological environment. Thus learning has become a life-long activity. We are being forced to face a basic truth that, in the past, was easier to ignore, namely, that life is itself a process of learning. Life means learning and growing. It has no final equilibrium, no plateau of rest. Life learns, adapts, and evolves. What is it that underlies all of these activities? What is learning, really?

8.3 Development and Learning

Every question about development and creativity that was raised in earlier chapters gives rise to an analogous question about learning. How does a person learn a new conceptual system? How does that new system manage to arise from a well-established older system from which the new system is incomprehensible? Does deep learning, the movement from one conceptual system or way of thinking to another, happen in a continuous way or does it proceed in discrete stages? If it is the latter then what are the mechanisms involved? If it involves both continuous and discrete mechanisms then is the way the brain and mind are employed in discontinuous learning identical to the way it is used in the continuous case; or is it the case that discontinuous learning requires a different way of using the mind? Grappling with these concerns will force us to confront some essential questions about teaching and learning.

We are interested in what I would call creative learning, and the mechanisms behind this kind of learning are identical to those involved in any significant creative act. The close parallels between learning, development, and creativity should force us to reevaluate many of our attitudes and practices in education at all levels. Most often it is felt that one cannot expect creativity from students, that creativity is only possible after many years of arduous preparation of one sort or another. However if that preparation, which is often what we mean when we think of education, includes developing fluency with the language and the concepts of a complex conceptual system, then the preparation already involves significant creative activity. Creativity is the essence of human learning at all levels, from the learning that is already going on in the womb to learning about the secrets of the natural world that is involved in scientific work.

Creative learning, the kind of learning involved in becoming fluent with a conceptual system or of going from one conceptual system to another, is the most important kind of learning. It is learning that sticks with you and is not quickly forgotten, that facilitates the growth of one's understanding. It is the learning that opens up new vistas for the individual that may be the basis for further exploration and originality.

8.4 Data and Knowledge

Learning is a dynamic activity and so the belief that learning is primarily about the acquisition of facts is fundamentally flawed—the acquisition and manipulation of data is at best a prerequisite to learning. Real learning involves acquiring knowledge and understanding. What are the characteristics of these distinct levels of learning: data, knowledge, understanding, and creativity?

Data is often confused with knowledge, and as a result, the words are often used interchangeably. There is, however, a difference between the two that is vital for my purposes and for education. Data appears to be purely objective—facts are facts, one often says, and the realist can only accept them. The whole value of a database is that the information it contains is assumed to be the same for everyone. Data seems to be objective, that is, it appears to be independent of any intelligent being who might interact with it. This is less true for knowledge, at least as I shall use the term, which must be knowable not by a computer but by a human being. Most people think of knowledge as standing on its own because the focus on "what is known" tends to obscure the cognitive dimension of knowledge. Knowledge must be known; it requires an actual or potential "knower."

A list of facts is not knowledge since knowledge must also involve patterns or relationships between individual facts. Patterns do not exist in the same objective way as facts. Relationships and patterns must be discerned, that is, they have a cognitive dimension. The Big Dipper is a pattern of stars but there is a difference between the stars and the Big Dipper. The stars are there all the time but someone must see the Big Dipper. Another example involves the problem of filling in the next numbers in the sequence: ***1***, ***3***, ***5***, ***7***, ***9***, …? The solution is, of course, that any number can be the next one. The sequence could continue ***11***, ***13***, ***15***, … or ***2***, ***4***, ***6***, ***8***, … or even ***0***, ***0***, ***0***, ***0***, …. Your answer will reveal what pattern *you* perceive. Patterns are personal. Nevertheless in certain situations there may be a social consensus and so most people will see the same pattern.

An amusing example of the difference between knowledge and information comes from the intelligent robot "Data" on the old Star Trek series. Intelligent robots have been featured in a series of movies but these movies probably say more about human hopes and fears than any realistic prediction of the future. "Data" was continually faced with human situations that he didn't quite understand, often involving non-verbal human capacities such as emotion or intuition. "Data" had access to a vast amount of information but he was continually reminded that something was missing—that he was not human.

These representations of intelligent computers are often romanticized and misleading but they do put their finger on an essential point, namely, the distinction between data and knowledge. Machines live in a universe of data; they may be intelligent but are not conscious. Knowledge, on the other hand, is necessarily linked to consciousness. It necessitates a human presence. Mathematics can be looked on as information—a list of every known theorem, for example. However mathematics is really a body of knowledge. "Doing" mathematics and learning mathematics involves acquiring knowledge.

Cognition is in this way analogous to perception. An act of perception, "I see the tree," has three ingredients: what is seen, "the tree"; the seer, "I"; and the "seeing," which relates the two. We sometimes forget that the primary ingredient here is the "seeing," which connects the objective pole "tree" to the subjective pole "I." Discussing the tree in the abstract obscures the "seer," the human dimension of seeing, but that dimension is present, if only implicitly. Similarly we might say, "two plus two equals four" but we should say, "I know that two plus two equals four." An act of cognition is the primary event that connects the subjective "knower" with objective "knowledge."

Data and knowledge differ to the extent to which the "knower" is acknowledged. Data feels totally objective, which actually means that the human dimension is more or less ignored. Knowledge also seems objective—we think of mathematics or quantum mechanics as objective bodies of knowledge and forget that human beings created them, and that they continue to exist only to the extent that they are of interest and value to various groups of people. Thus the knower is not as well hidden in the case of knowledge as it is in the case of data. Nevertheless there is a tendency in the modern world to reduce all knowledge to data and this is done by completely suppressing the "knower."

One task of modern education is to make the crucial distinction between data and knowledge by encouraging conceptual development, which alone can give significance to information. Knowledge differs from raw data because it is conceptual—concepts have a cognitive dimension. A formal definition is not knowledge. Concepts must be knowable and knowledge is built out of concepts grouped together into conceptual systems. A concept is not a fact; it is not cut and dried.

Thus knowledge lives at a "higher" or more sophisticated level than data or facts, where the word "higher" indicates the extent to which the cognitive dimension is acknowledged. The facts of the multiplication table do not automatically confer knowledge of (the concept of) multiplication. The "facts" won't tell you that multiplication can be represented by area or by repeated addition. They won't necessarily reveal the mathematical properties of multiplication such as commutativity ('***a***' times '***b***' equals '***b***' times '***a***'). Knowledge of multiplication carries some implication of being able to work with multiplication and being able to discern some of its properties.

DNA research gives a good illustration of the distinction between data and knowledge. Sequencing DNA involves reducing it to data. This is an important accomplishment but only a preliminary one. It is true that this enables one to solve some important problems. However it is amusing to hear some of the excessive claims that are made on the basis of this research. In fact we are only at the beginning of an enormous project, for the real work involves deciphering the lengthy DNA sequence. For example, today we are becoming aware of the crucial role of epigenetic and other factors which determine when (or if) a given gene is "turned on." Thus, for many situations the raw DNA code is insufficient to allow us to understand the nature of complex problems like cancer or mental health. There is a great deal of work that is required to convert DNA data into actual knowledge. This is where the breakthroughs will arise. The data is preliminary work.

8.5 Understanding

For knowledge to be useful it must be understood. Whereas knowledge is primarily social, understanding is individual—it is tied to a particular individual. It makes no sense to talk about understanding independent of the person who understands. One can talk about "understanding multiplication" but one person's understanding may well differ from another's. There is no template for proper "understanding" that is the same for all people at all times. Knowledge only comes alive when it is understood.

I have even had the experience of looking at a research paper, that I myself had written many years ago, and not understanding it. In order to understand it I would have had to put in the considerable effort of immersing myself again in the concepts and ideas that the paper dealt with. In other words understanding is not some phenomenon that never changes. Understanding is a cognitive phenomenon; it is a kind of insight into things. You cannot point to it as you would to some objective phenomenon and say, "this is understanding." Understanding can grow or diminish. Take some complicated mathematical concept like randomness or continuity. At one point you may say, "I understand randomness." But at a later date you may turn around and say, "You know I never really understood randomness before. Now I understand it much better." Furthermore understanding something complex like randomness, or learning and thinking for that matter, is a process with no end. You can always understand it better or from a different angle. It is always possible to rethink your understanding in the light of new evidence. In the language I used earlier, it is always possible to reframe one's understanding.

What we look for in education is mostly a minimal level of understanding that almost everyone can attain. Even so teachers often forget that understanding is individual—different people can legitimately think about multiplication in different ways. You cannot get into someone else's head and verify their understanding and you cannot impose your understanding on other people. This is why teaching is so hard and good teachers are rare. I shall have much more to say about this later but for now it will suffice to say that learning is about gaining understanding.

Since understanding is always conceptual it makes no sense to speak of understanding data. It is only possible to understand knowledge and you do this by making the knowledge your own. Moreover understanding, as the term will be used in this book, usually requires more than knowledge of individual concepts. It requires seeing the relationship between different instances of the same concept or between different concepts. Understanding multiplication includes many things including a familiarity with its various representations and its relationship with addition and division. It involves viewing multiplication through various conceptual systems—the counting numbers, the fractions, the real or complex numbers.

Facts and concepts only acquire real meaning and significance when viewed through the lens of a conceptual system. However the conceptual universe is not permanent but, on the contrary, is subject to continual revision and even to radical change. It is not even really objective since it varies with the individual; nor is it merely subjective since it has properties for which there is a social consensus. Learning and understanding means entering into this consensus and developing one's own unique take on it.

Nevertheless many persist in thinking that education involves acquiring objective and timeless knowledge. The fact of the matter is that data on its own is meaningless even if it seems to be objective and permanent. As soon as the human being enters the picture, and that happens when you discuss concepts and conceptual systems, you have objective and subjective dimensions, which need to be taken into account.

Facts do not exist independently of knowledge and understanding for without some conceptual basis one would not know what data to even consider. The very act of choosing implies some knowledge. One could say that data, knowledge, and understanding are different ways of describing the same situation depending on the type of human involvement implied—"data" means a de-emphasis on the human dimension whereas "understanding" highlights it. Of course the top of the hierarchy of learning is creative learning, which, to be consistent, is present at all the other levels but, unfortunately, is often ignored. All of the errors of education involve ignoring this hierarchy, which happens when one views learning and teaching as the mere transfer of information.

I shall look at facts, concepts, and conceptual systems as temporary equilibriums—stages—that inevitably arise within the learning process. We tend to focus on these equilibriums and to imagine that they are permanent instead of focusing on the way that they arise, break down, and morph from one into another. If you want to see learning in action then look at young children, artists and scientists. Look at the process of evolution. Everywhere you will see the same dynamic playing itself out; everywhere you will see learning.

8.6 The Reality of Change

Change is real; the unchanging is an illusion. Classical science focused on systems that were in equilibrium whereas modern science also looks at states that may be far from equilibrium. Equilibrium situations can give you the feeling that things are unchanging but this is always only a temporary condition. Equilibriums beak down and when they do, when the system is far from equilibrium, then the dynamism of the system is most visible. In such situations one is studying change directly.

The theory of evolution is, of course, a theory of change and its moral is that life and change are inseparable. However the evolutionary point of view does not only apply to living systems but also to other domains—from the physical universe of planets, stars, and galaxies to the circuits in our brain, which evolve in tandem with the development of psychological and cognitive states. Evolution studies the rules that govern change and so should any discussion of learning.

The reason that change has often been ignored in classical science might have been that certain changes are positively glacial in comparison with the human life span. At one time, not so long ago, people's lives were pretty much identical to that of their parents and grandparents—the changes that were visible were mostly the cycles of nature. This is the reason that it is so difficult for human societies to deal with environmental changes like global warming—heredity has only equipped us to deal with emergencies that take place over brief intervals of time but not with developments that take decades or longer to unravel. On the other hand, contemporary society is characterized by the accelerating rate of change; changes that are primarily initiated by advances in technology.

It is easy to point to the negative effects of rapid change on individuals and society and many commentators have done so. There is nowhere to hide from a job that follows you twenty-four hours a day on your smart phone. It can make life very stressful. There is no oasis from society, advertising, and even friends—no peace, tranquility, or solitude. Surely it is important that people today learn to carve out some time and place where they can shut out the demands of the world and turn inward to refresh and center themselves.

Nevertheless the demands of change can be framed in another, more positive manner. Change is difficult precisely because it is so real and thus defeats our attempts to create an illusion of stasis. For long stretches of our lives we can maintain the illusion that we have not changed either physically or psychologically. Yet one day we wake up to the fact that ten, twenty, thirty, or more years have slipped by and we are not the person that we were. The challenge is to learn to live with the reality of change and to live every moment as completely as we can. Today it is more difficult to maintain the illusion that we, and the world, are unchanging—we see change with our own eyes, read about it, and listen to intelligent observers discuss it—yet we do little about it. The challenge for the educational system and for society as a whole is to adapt institutionally and socially to the omnipresence of change, not to some particular change, as was the case in the past, but to the reality of continual change.

8.7 Learning is a Positive Response to Change

Learning is a positive response to change in which one openly invites change into one's life. None of us are experts at facing change. We all have something to learn in this regard and both sides of the educational divide, teachers and students, may well have something to contribute. It may well be that a six-year old child has an attitude towards learning that the so-called experts could do well to emulate.

The function of modern education should be to prepare students to prosper in this world of continual change. Our cognitive structures are always playing catch-up with the world. Just when we thought that we could relax because we have acquired the "right way" to look at things, we find that things have changed and the "right way" has become the "old way." We then have to master new concepts and new conceptual systems, not to speak of giving up our old ways, which have now become obsolete. This is the challenge of modern education. Everything and everyone is changing all the time and education must become an entry into this world. How do we learn to balance our need for stability and security with our need to live in the world of change, which is stressful but also exciting and stimulating?

8.8 Learning as the Transfer of Information

Let me return to the relationship between learning and the transfer of information. What is learning? When I look up "learning" in the dictionary the first thing I find is "the acquisition of knowledge or skill" where "knowledge," is then defined to be "the awareness of information or facts." This is clearly not the way in which I have been talking about either learning or knowledge. For me knowledge is conceptual. It is only by ignoring the conceptual dimension that learning becomes the acquisition of information.

This attempt to frame education as the mere transfer of information is a powerful tendency in the modern world. From this viewpoint, the teacher, book, or computer program has the information and transfers it into the head of the student. This approach towards learning does not work in practice because it is wrong in many different ways. It may refer to the data acquisition of computer systems but not to human learning. The human mind has a lot of trouble with even remembering random facts. In order to remember something that is at all complex—like a mathematical argument—it must be understood. It must have meaning and that meaning is carried by a system of ideas. Learning involves mastering ideas and ideas are not facts, they are relationships.

Our age has often been called the "information age" and so it is natural that we identify information with knowledge and think of this knowledge as facts. We focus on the material to be learned and not on the conceptual system by means of which we give meaning to the material. We believe that information is neutral and stands on its own and that the educational task primarily involves mastering large amounts of information. The point is that we emphasize "what we learn" and de-emphasize "the learning process," the cognitive dimension of education. As a result of this mistaken approach learning can easily be reduced to rote learning, a kind of learning that has its place but is far from the creative learning that this book is about.

Thinking of learning as data acquisition is the direction in which much of modern education is oriented. Educational institutions have been compared to factories that process raw materials—facts—and convert the raw materials into a finished product—the educated person who knows the relevant facts and their conventional uses. It is based on a way of looking at the world that arose with the industrial revolution but is today outmoded and inappropriate to the age in which we live.

This erroneous orientation is the main reason why our educational institutions have failed for the most part in what should be their main mission—producing a public that can critically evaluate, and adapt to, the rapid changes that characterize the modern world. Education is, or should be, a preparation for the real world and the acquisition of knowledge that will soon be obsolete is not the training that is needed. We have an educational system that is a carry-over from another time. When we think of fixing it we gravitate towards things like information delivery over the internet or other kinds of computer assisted education. This looks like it is a new approach but it is really a very old approach dressed up in new technological clothes. We are still in the old conceptual system when what we need is a new one that is appropriate to our time. But we first have to identify the problems with the old before there is space for a consideration of the new.

8.9 Learning as the Acquisition of Skills

Sometimes the discussion of learning expands beyond the accumulation of facts to the acquisition of skills. Education certainly involves learning skills: how to write an essay, program a computer, or differentiate a function. The learning of even basic skills like language and arithmetic is at a higher level than mere transfer of information because it is open-ended and contains an element of creativity. A child who learns a language can formulate sentences that they have never heard before. Children who have learned the counting numbers can answer questions like, "Is there a largest number?" Or they can make observations like "if 2 plus 3 equals 5 then 20 plus 30 must equal 50." Expressing a new sentence or recognizing a new arithmetical regularity are insights that demonstrate that the child has acquired a certain knowledge of language or of the system of counting numbers. In both language and arithmetic, "facts" are replaced by insights into conceptual patterns—which a child learns to recognize and then apply appropriately in new situations. There is something creative in a child's mastery of language and arithmetic.

The simplest and most frequently learned skills involve the systematic application of rules—algorithmic learning. The usual addition or multiplication algorithms are good examples but algorithmic learning is not limited to mathematics. Yet mathematics is a good example. At one time a good part of a student's education was spent mastering standard algorithms but today this has mostly been rendered obsolete (for good and for bad) by computing devices. When I talk of learning a skill I am usually thinking of something that is algorithmic. Learning to write an essay, for example, involves various skills that can be learned in this way: skills such as spelling, grammar, punctuation, sentence and paragraph structure, and so on. Of course writing a good essay goes beyond all of these particular skills. It's really an art form and so has many aspects that cannot be learned algorithmically. Thus writing involves learning skills but also involves another dimension—the creative.

8.10 Learning Concepts and Conceptual Systems

I think of the acquisition of data and skills not as real education, but as a preparation for real education, a necessary prerequisite, perhaps, but still only a prerequisite. Real education involves the acquisition of knowledge and understanding, which both include an important element of creativity. A crucial step involves concept formation. In mathematics and physics one needs to grasp concepts such as mass, velocity, irrational number, or randomness. Mastering a concept requires creative work. It is not enough to memorize a formal definition. You have to know what the definition *means* and this implies having a "feeling" for the concept. You have to have experienced lots of concrete examples of the concept and then make the leap to what all these examples have in common. And finally you have to be able to apply the concept appropriately, even in situations that are new.

As we have seen even elementary arithmetic involves concepts like number and the arithmetical operations. Even these elementary concepts are very subtle as they involve both information and procedures. Grasping these concepts involves rote and algorithmic learning as well as something else—a creative X-factor. You can call it conceptual learning or conceptual understanding. It involves deep thinking and so it arrives suddenly as a result of overcoming some obstacle. It often involves the unlearning of an earlier concept that has proved to be inadequate to the task at hand. The new concept typically comes with multiple representations. Understanding, what the mathematics educator Anna Sierpinska calls "good understanding" in her excellent book

Understanding in Mathematics, involves seeing these different representations as different views of a unitary idea.

Nevertheless, individual concepts are not enough. They must be integrated into a conceptual system. In one way the conceptual system is at a higher level than individual concepts; although, in another, individual concepts get their meaning through conceptual systems. Thus it is a mistake to think of individual concepts in isolation from one another.

8.11 Reframing: Learning as the Change of Conceptual System

Finally, and crucially, education involves replacing one conceptual system by another. This last step is the key, and most difficult, aspect of education but this is what the modern world calls on us all to do on a regular basis. This is the step that could be called "learning how to learn." Learning is not just about acquiring knowledge, skill, concepts, or even conceptual systems. It is also about moving from one way of thinking about a situation to another, more complex, way of thinking. This latter step involves acquiring something—the new system—but also pruning away something else—the total reliance on the old system in situations in which it is inadequate. Normally learning is only considered from the point of view of "gaining something" but it also involves "losing something." Growing involves letting go of what has been outgrown and this can be painful and difficult. It is at this stage that the full complexity of the educational task is revealed. It is here that our discussion of creativity, in all of its subtlety and complexity, is relevant. This is the primary activity that I have in mind when I use the expression "creative learning."

8.12 The Spectrum of Learning and Creativity

To sum up, we have seen that it is not possible to characterize learning with precision because learning comes in many different modalities: the transfer of information, algorithmic learning, concept formation, the development of conceptual systems, and, finally, reframing or changing one's operational conceptual system. We should really talk about the "spectrum of learning" because the different aspects of learning that I have enumerated come with a direction. The spectrum moves in the direction of increasing complexity. As we move from the first to the last we see an increase in creativity. In fact it might be reasonable to add on one final step to our spectrum and that would be what we normally consider to be creative work—the development of new scientific theories, works of art, or technological innovations. Normally creativity is only associated with the final step, with the production of something that is new and has never been seen before. Looked on in that way we forget that the creative process itself is not new and has always been around. When we come to see learning as a spectrum we will also see creativity in the same way. When all learning is seen to contain elements of creativity then it cannot help but transform our view of what education should be.

8.13 Knowing

Every successful learning situation involves knowing but the word knowing can easily be misunderstood. Sometimes we know *something* for example, a fact or a procedure or maybe a concept with the proviso that knowing a concept is not the same thing as knowing a fact. Mastering a conceptual system also involves a form of knowing but this is not so much a knowing of something as of "knowing through something." This was the point that I was making through the use of the metaphor of a conceptual system as a window.

It is interesting that if we could divorce the object of knowing from the feeling of knowing itself—the transitive versus the intransitive senses of knowing—we would find that it is the same "knowing" that we feel in all of these situations. The result of learning is knowing but we tend to focus on something else—*what we know* as compared with *that we know*. This comes from thinking of learning as analogous to an industrial process. When you make widgets you worry about the kinds of widgets, their quality, and so on. The process is just there to produce widgets—it is secondary. But learning is not an industrial process and the industrial metaphor is a bad one. Learning how to learn, and the joy of learning, comes from accessing the domain of knowing. It is invariably associated with acts of creativity. It is its own reward and is the reason that learning feels so good. When people used to do Euclidean geometry in school they would get that feeling whenever they saw into the resolution of a deduction. When you replace this kind of problem by one whose solution depends on an algorithm—differentiating a function, for example, or multiplying two five digit numbers, you don't get this release. There is no "aha!" moment. Nothing has been learned.

8.14 The Problem with Education

Our educational systems are in disarray because they have been set up to do the wrong job. We have not defined the task of education properly and, as a result, we cannot possibly succeed. For the most part we have given up on the real task of education—on deep learning and the development of deep learners. Many teachers may have heard that education is a creative process, they may even believe that it is in their hearts, but few approach it in that way in the classroom or university lecture hall. There are many reasons for this. For the most part they would not know how to do so because they have not have been educated in that way themselves.

Even for those teachers who are in touch with the creative sources of learning or who are creative in other areas of their lives find that it is just too hard to teach in this way. It is hard for the obvious reasons—large class sizes, problems of discipline, lack of preparation and motivation on the part of the students, and the distractions of electronic gadgets, the internet, and the media. However the deeper reason why nurturing deep learning is hard to do is that creative learning involves the kind of difficulty I identified as one of the characteristics of deep thinking. The difficulty of learning is qualitatively similar to the difficulty of producing a great work of art or a new scientific theory. How do we orient our educational system in this direction? This is the key question.

At the university creative activity is identified with doing research. It is assumed that students put their creativity on the shelf through their undergraduate and a good deal of their graduate career. At a certain moment we demand a magical metamorphosis in students. They were merely amassers of information but are now expected to be creative researchers—to think, not in terms of mastering the old, but of producing the new. Then the students are surprised and their advisors are disappointed when the change comes slowly or not at all. If education were approached in the manner that I suggest in these pages then the gaps between undergraduate education, graduate education, post-doctoral training, and full-fledged research activity would be seen as the continuum that it ideally should be—the same creative activity, the same basic process, just pursued in different environments with different degrees of autonomy.

Let us consider the essence of education—the movement from CS_1 to CS_2—from the point of view of neuroscience, for a moment. Then a conceptual system would correspond to a circuit or set of circuits in the brain. A conceptual system that is understood and used repeatedly corresponds to circuits that are well entrenched. The problem of changing from one conceptual system to another would involve establishing new brain circuits. Since the brain obeys the law of "use it or lose it" we can see that for a considerable time there will be a tendency to use the old, well-established, circuitry. How do we set up the new circuits? What is the connection between the new circuits and the old? How do we keep ourselves from continuing to use the old in preference to the new? These are not easy questions but implicitly or explicitly education demands operational answers to them.

Accomplishing this kind of fundamental mental and behavioral change is difficult and stressful and so the student needs help and support. He or she cannot hope to get that kind of support from someone who is unaware that this is what education requires. Maybe it is years since the teacher has thought about mathematics in a new way; maybe she has totally lost touch with a creative or even a playful approach to mathematics. Maybe it has just become "stuff," that is, information, to be "downloaded" from the text to the student. If so it will be impossible for the teacher to comprehend the situation in which the student finds herself much less to provide the guidance and support that she needs.

The sad truth is that we often don't try very hard to educate students and are willing to accept a facsimile of education. This is a great tragedy—for the teacher, the student, and society as a whole. Society needs people who can think and innovate. Both students and teachers are human beings with their own basic needs for creative expression. Schools, colleges, and universities are ideal places in which to learn how to tap into one's inner creative resources. The facts will be quickly forgotten; algorithms will give way to improved computer programs; but learning how to learn and to approach life in a creative manner is a priceless gift that is transferable from one situation to another. Teaching in this way changes the teacher; learning in this way changes the student. Long after the facts are all forgotten, these changes will remain alive and well.

8.15 There is No Learning without Obstacles

Learning does not just happen; it is the result of solving problems and overcoming obstacles. When we think about education as the acquisition of information and technique it is as though we were picking flowers from our garden—it may take a little time and energy but there is nothing that obstructs you. It's not the same when a student tries to understand some topic in mathematics. She can spend hours of her time without necessarily making any progress. One of the hallmarks of creative activity is that it is undertaken without any guarantee of success because it is based on encountering and overcoming obstacles of various sorts. Let me repeat what I said earlier, namely, that all learning, whether as development, school learning, or scientific research, is a discontinuous process, which starts with encountering an obstacle that blocks future progress until such time as it is overcome. The existence of obstacles is an essential feature of education, which is often ignored. You can't think sensibly about education without seeing it as a response to the problematic. That is what's wrong with the information theoretical approach—because it starts out in the wrong direction it will never get anywhere interesting.

Education is not a process of gathering and sorting information like a Google search algorithm (brilliant as that may be)! Education—creative learning—is an adventure into the unknown. It takes courage to face a situation that is new and makes no sense to you and yet that is what we regularly ask students to do. In comparison, we adults often do our best to avoid situations that are challenging in this way because they make us uncomfortable. Often we equate the good life with the comfortable life. When you face something that you don't understand it is evident that you are not in control of the situation. It is then tempting to succumb to methods such as algorithms that appear to guarantee success. Unfortunately this amounts to replacing the unknown by the known, by the tried and true. It is like trying to use logic to deduce a new idea—that is just not the way it works. The great mathematician Andrew Wiles was asked, "What do you do when you hit a brick wall?" He answered simply, "I would go out for a walk. I'd walk down by the lake. Walking has a good effect in that you are in a state of relaxation, but at the same time you're allowing the sub-conscious to work on you." This is more like how you open up a space so that there is room for a new idea to come bubbling up. It clarifies the necessary elements of both education and creativity—obstacles, absorption, and relaxation—but the whole process cannot even get started if one does not face up to the challenge of the problematic.

Is it possible to learn to work with problematic situations, to learn to enjoy them? As young children we were all avid learners and faced up to all kinds of obstacles on a regular basis. We were not afraid to take a chance. What happened to most of us as we grew up?

8.16 Learning is Humiliating

A short anecdote will illustrate the problem. I was intending to discuss creativity in a class that I was teaching to some students in an interdisciplinary Liberal Arts course. In the previous class I had assigned an essay that the students were supposed to read and comment on. In the next class I was informed in no uncertain terms by some of the students that they had found this particular essay very difficult to understand. Now mathematics students regularly encounter new ideas that they initially find difficult to understand. These students were not used to being in such a situation and didn't like it at all. Many of them came to class angry—at me and at the writer of the article. They said things like, "I don't understand what he is saying. Surely he could have written it in a simpler way. Why does he have to use such big words?" The virulence of their reaction was not proportional to what was at stake in this situation. I took it as a hint that something more general and important was going on.

It was not merely that they did not understand this particular essay but that they felt humiliated by the situation. Humiliation is a strong word but could it be that, for some people, incomprehension is experienced as humiliation? Perhaps some students experience learning situations as humiliating. This may be because they have had a bad experience in the past where a teacher or fellow student held them up to ridicule or because they themselves have doubts about their own capacities. But there is also another possible reason. Learning inevitably begins by admitting your own ignorance, even if this is only to yourself, and this makes you feel vulnerable. It is this vulnerability, and the impossibility of protecting oneself from it within the learning situation, that can be experienced as humiliation. We spend a good deal of our lives protecting ourselves from feelings of vulnerability by being clever or aggressive or funny or cute. All of these responses keep us from even seeing the learning obstacle much less overcoming it.

If this is true then the educator is in a double bind. If you teach conscientiously and so expose students to the obstacles that are always present in real learning then you run the risk of humiliating them. If you allow the students to pretend that there are no obstacles or that they can be avoided then you are not really teaching. The threat of humiliation is never far from any learning situation—you can see it in the faces of the students as they attempt to avoid eye contact with the teacher. The question is what do we do about it? I'll come back to this later on.

I asked these particular students, "Surely this is not the first time that you have encountered something that you did not understand at first reading? What is your strategy in such situations? What did you do when you found that you did not understand this essay?" There was an uncomfortable silence in the room, which I did not interrupt for many minutes. No one, it seemed, had a strategy. One student finally said rather weakly, "Read it again?" Mostly they seemed to feel that introducing difficult material was breaking some rule of the game, that it was the teacher's job to digest the material for them. Learning in this view is a passive activity like watching a TV program with the teacher as entertainer. You can get information out of a TV program but you don't learn anything important from it because you are not challenged to change and grow. The education that these students had undergone had mostly avoided obstacles—the kind of education that I am advocating goes in the opposite direction.

8.17 Not Understanding

There is an observation that is apropos here that I am almost embarrassed to make because it is so trite and obvious. It is that you have to "not understand" *before* you can hope to understand. If you can never accept being in the position of "not understanding" then you will never understand. Let's think about "not understanding" for a moment. Because it is phrased negatively we think of "not understanding" merely as a lack, a hole that we are called upon to fill up. Not understanding makes us uncomfortable and we think that the way to go is to get rid of the discomfort as quickly as possible. We don't focus on what it feels like to "not understand." It is a valid sensation in its own right, which is where any learning experience originates. "Not understanding" is not necessarily painful or difficult. It is blocking it out of your consciousness that provokes difficulty because it puts you in a double bind—you don't want to own up to it because it makes you feel inadequate and out of control but you have to admit it if you are to make any headway in learning the new idea.

Actually "not getting it" can be a most powerful and pleasurable experience. Suppose you are looking at a beautiful sunset at the end of a long summer's day and you are swept up in the sense of awe that it evokes. Everything feels perfect but if you try to analyze and understand that perfection it just falls away. The essence of this feeling is precisely "not understanding"—an openness that in this case is sufficient in itself. Everything, in that moment, is all right. At its best, learning asks us to open up in this way.

When we expose ourselves to a new learning situation we must begin by accepting where we are without judgment and where we are is that we don't get it. Not understanding does not mean that you are stupid or lacking in some way. It is just part of the process.

Actually not understanding is quite subtle. Examined carefully, one finds that it changes over time and, at a certain point, from within one's absorption with this vaguely defined situation, there arises the first glimmering that something systematic is going on even if, at that moment, you can't yet articulate what that something is. Understanding happens suddenly but the insight is preceded by preliminary hints. Patience and confidence is required to bring these initial hints to fruition. However the entire process is aborted if the individual is unwilling or unable to face the situation of not understanding. This is the primary obstacle to learning anything new.

8.18 Epistemological Obstacles

Every learning situation has a whole collection of possible obstacles and there is a vast literature in educational research cataloging the obstacles that are associated with learning a given concept or conceptual system. Thinking about obstacles as opportunities brings me to the idea, introduced by the French physicist and philosopher of science, Gaston Bachelard, of "epistemological obstacles."[i] In her book Anna Sierpinska writes, "He (Bachelard) found that students' thinking appeared to suffer from certain 'epistemological' obstacles that had to be overcome if a new concept was to be developed. These 'epistemological obstacles' [were] ways of understanding based on some unconscious, culturally acquired schemes of thought and unquestioned beliefs about the nature of mathematics and fundamental categories such as number, space, cause, chance, infinity [that were] inadequate with respect to modern day theory

[but] marked the development of the concept in history. For example students do not understand "***.999... = 1***" because they do not grasp the concept of completed infinity and, in that, they are merely repeating the difficulties that historically faced the mathematical community as a whole."

This notion of obstacles as epistemological was a powerful insight and Sierpinska reports that it generated a great deal of research in mathematics education where one searched for key epistemological obstacles in various learning situations. The notion is important to my thesis for two reasons. First of all it underlines the view that conceptual development is concerned with meeting and overcoming obstacles. Second, it highlights the parallels between individual learning and the historical progress of science and mathematics, which provides evidence that supports my generalized hypothesis about the existence of parallels within all areas of development, creativity, and learning. The orientation of my thesis also parallels Bachelard's argument with the neo-positivists of his time. Learning is not algorithmic; mind is not algorithmic. There is something else going on, something that is essential, but is very subtle and mysterious. The mathematics educators Dubinsky and Levin claim, "It would seem that one never has direct access to cognitive processes…but, at best, only to what an individual can articulate or demonstrate at the moment of insight itself. Precisely what occurs at that moment seems as inaccessible as it is essential."[ii]

The moment of insight is the moment in which one grasps the concept or makes the creative leap from one conceptual system to another. It is the essential ingredient in deep learning, something that has a good deal in common with Sierpinska's notion of "good understanding." It is present in every act of creativity. It is the factor that brings education to life. It seems mysterious to us for the obvious reason that it cannot be pinned down or reasoned out. It transcends our analytic intelligence—it is a right- not a left-hemispheric phenomenon.

We are not even conscious of the moment of insight but, as Dubinsky and Levin discovered, we can only be conscious of it after the fact. You might conclude that insight is difficult and complicated but, on the contrary, it is so natural that even a baby can do it. We have difficulty understanding creativity because we have become too complicated.

8.19 Unstoppable Learning

Sujata Mitra is a Professor of Educational Technology at the University of Newcastle. He performed a fascinating experiment, which tells us a lot about the nature of learning.[iii] Mitra placed computers in a wall in various slums and small villages in rural India. Then he left the scene and monitored what happened using a remote desktop. These computers quickly attracted the attention of the children of the area. Unfortunately the children couldn't easily use the computers because there were no operating instructions or teachers, and anyhow the computers used English, which was a language that the children were not familiar with. They would have to teach themselves English in order to use the machines.

And they did! Not only did they learn the English that they needed but also the computer literacy skills that were necessary. This was so unexpected, so extraordinary, that Mitra believes that he has uncovered some universal principle of great significance. "Maybe," he suggests, "learning can happen on its own." "Learning is a product of educational self-organization. Let it happen." The role of the teacher is to set it in motion, which he did in this case by planting those computers in a conspicuous place. The children did the rest.

This idea that learning is unstoppable, the natural way that the mind functions, has come up repeatedly in my discussion of the basic properties of deep thinking in various contexts. Consider, again, the development of an infant. As I said earlier, the child does not decide to learn; development *is* learning. Every child's formative years consists of an intense process of learning. What amazing discoveries children make in those early years: they learn to discern individual objects; they learn order and number; they learn language, they learn to walk; they create a self. Each one of these accomplishments is extraordinary. We take them so much for granted that we don't stop to consider how extraordinary they are. They are at the same level as the great creative leaps taken by our very greatest artists and scientists. Yet we all manage them—we all learn to walk and to talk. So we are all extraordinary—every child is the heir to this creative potential.

What is so striking about infants is that they are basically little learning beings—continuously participating in what is happening at the moment, continuously attempting to map out their relationship with their immediate surroundings. They are young scientists systematically exploring their world, a world that is simultaneously being created by them. Too many people believe that children are deficient in some way and that our job as adults is to make them exactly like us. But children, especially infants, possess an intelligence that adults don't always appreciate. They have as much to teach us as we have to teach them.

But learning is not just for infants. The basic drive behind scientific research is the need to understand which is actualized through the development of appropriate conceptual structures; in other words, research is learning the right way to think about the world. Thus there is a natural parallel between the development of science and the development of children as remarkable as that initially seems. Moreover one finds the same kind of developmental learning is the essence of all life. After all what is evolution if it is not a history of adaptation, growth and development and what does adaptation mean if not learning? Thus learning is central to life, and, as a result we are attuned to learning at the most basic levels of our biology. DNA is itself a mechanism for learning and continual change.

Moving to another level, which is more basic and may seem a little far-fetched to some, the evolution of the physical universe as revealed to us in modern physics and cosmology can itself be seen as a form of learning. This will give us a different way of thinking about cosmic history. Electrons and protons learned how to combine into the singular structure of a hydrogen atom. Hydrogen and oxygen learned how to get together to make water. Certain proteins learned how to combine to form cells. That the history of the cosmos is a history of such leaps is not in question. Most people would not use the word learning in this context but thinking about the analogies between all of these radically different situations (which I emphasize by calling them all "learning") will give us insights into both the general situation—what learning is—and specific situations like child development, evolution, or the history of science.

Creative Learning is basic to all life and to the cosmos itself. One might say, "Learning *is* being." To learn, to grow, to evolve is to be alive—it is the essential characteristic of life. Life is the flowering of a thousand possibilities. It is constant change and has been so from the very beginning, from the big bang and the creation of time, space, energy, and matter. Everything that we know and experience came into existence. In the course of this ongoing creativity lessons were learned and structures attained some stability. We can discern the mechanisms of learning in all the sciences from physics and cosmology to biology and psychology.

Normally we think of learning as an outcome of consciousness—we think of it as an outcome of thought and experience. However creative learning, the way it is described in this book, is more basic than that. In fact we really should turn things around. Consciousness itself is learned and arose historically at a definite stage in the ongoing process of evolution. It is hard to imagine thinking without language even if this language is not made up of words but is musical or mathematical. However words, notes, and numbers were all created at a given moment of history possibly based on some innate cognitive structures that we are born with. In other words these words could not be written; these ideas could not be expressed; if it were not for a learning process that stretches back to the beginnings of time.

Nevertheless words and numbers are still learned and refined by each individual in every generation. They need to be turned into "my words" or "my numbers." They will always retain those two dimensions—call them the personal and the social. The social dimension often appears to be objective and so people think that learning means grasping the social consensus, which often feels like an objective reality. They don't realize that it is necessary to get the meaning of the words or, even better, the meaning *behind* the words. Meaning always includes a subjective element. Learning spans the objective and the subjective; it involves objects and processes.

8.20 What is Teaching?

One question that arises immediately is this, "If learning is indeed universal then what is the role of the teacher?" My answer is the same as the one Mitra obtained in the experiment mentioned earlier. There is no need for a teacher, in the usual sense of someone stuffing knowledge into empty heads. The teacher is actually essential, as I shall discuss in Chapter 9. But the role of the teacher needs to be reexamined.

Similarly there is no need for a being or force that stands outside of nature and drives the evolutionary process or any other learning situation. Creativity is built into the natural world as it is into human beings. We must rethink the relationship between creativity and the natural world as well as the relationship between creativity and analytic intelligence. A reversal is required: creativity comes first, it is extremely general and basic.

There is no need to restrict what we mean by learning to the mere transfer of information from teacher to student. We have to redefine what we mean by "teaching" and "teacher." Like the child learning to walk we are all driven to learn by forces beyond our conscious control. If anything is the "teacher" it is this drive to learn. To make this even more dramatic we could say that we are all basically an expression of this drive to learn—that it is the most basic element in our make-up. Those who aid us on this path, those who help us to liberate our natural propensity to learn, we call our teachers. This is a subtle and rewarding role and it is not restricted to the classroom or the university lecture hall. I have learned some very profound lessons from carefully observing babies and interacting with children and students. When you are ready and receptive anyone can become your teacher.

I said earlier that there is little difference between learning and creativity and so if we are all driven to learn then we are all basically creative. Creative learning is not so much something that we do as it is something that happens through us—an opening up to something larger than our individual selves. Of course we can't just sit around waiting for creativity to strike. We must actively engage with the world around us. Learning and creativity are an expression of our unity with the dynamic forces of life itself.

Learning is natural and inevitable. There is really no stasis, no time when we are in a state of permanent equilibrium. Either we are growing and learning, or we are in a state of decline and decay. Learning unfolds within and outside of us in a natural way. It is true that we can impede it or try to force it into some preconceived direction but this distorts the process. Learning unfolds naturally within us yet this does not mean that it does not take any effort. On the contrary learning takes great effort but, as we saw in our discussion of creativity, it is a very subtle kind of effort.

When you are learning something but don't yet know it, the state that you are aiming for is unknown and unimaginable. When learning to count the child may begin by watching Sesame Street and repeating by rote the "counting numbers": one, two, three, four, and so on; first up to ten and later up to twenty. They memorize the list but they still don't know it's significance—they don't know how to count. Yet repeating the counting numbers, playing various number games, and watching adults perform various counting activities bring *all* children to the stage when they create an internal system—the conceptual system of counting—and abruptly enter into an entirely new world of counting, adding, subtracting, and multiplying of whole numbers. They have acquired an essential aspect of culture; they have given meaning to "number." When this incredible leap occurs you know that there are the same number of peanuts in a collection of ten peanuts as there are marbles in a collection of ten marbles. You can respond to the question, "What is the largest number?" by saying, "There is none; they go on forever."

Thinking about learning as an event is key to the view that is developed in these pages because, as Einstein taught us in the Theory of Relativity, the elements of the world are events. The world is characterized by change, by process. Events, unlike objects, happen, that is, they are dynamic so a world of events is suited to a description of a reality that is itself dynamic. Look at children, artists and scientists, and at the evolution of life and of the universe itself. Everywhere you will see the same thing going on, everywhere you will see the dynamism of learning.

When a child makes the great leap that we prosaically call learning the counting numbers, the same fundamental process is at play as when the first aquatic life adapted to life on dry land. Similarly a child who utters her first word has recapitulated the initial discovery of language. It is all learning and whenever it happens it contains elements of the miraculous. Every time that happens there is a shock of recognition, a feeling of coming home, of attaining a new stability, an oasis of protection from the threat of impermanence.

---

[i] Sierpinska (1994), pp. 133–135.

[ii] Sierpinska (1994), p. 120.

[iii] Mitra won the 2013 TED prize for the talk that can be seen on YouTube at: www.youtube.com/watch?v=HE5GX3U3BYQ.

Chapter 9

Creative Teaching

9.1 Introduction

It is scarcely possible to talk about learning without also discussing teaching and so I shall devote this chapter to a discussion of teaching. What is good teaching? It is very simple. Good teaching involves activities that enable creative learning. This kind of teaching is a creative activity in its own right. In order to nurture creative learning, the teacher must be aware of the dynamics of the learning process in others and in themselves. Every subject, every class, and every student is a challenge for such a teacher, a challenge that calls for a creative resolution. Every insight into the subject on the part of the student is accompanied by an insight into the student's understanding on the part of the teacher.

We discovered in Chapter 8 that learning is an innate force which is natural and cannot be stopped but can be blocked or diverted in directions which are not constructive. Thus one could say that teaching is secondary, even if this is a little simplistic. The need to learn, and the compulsion to do so, are universal. Without this drive, life would not survive and the universe would not be the dynamic environment that it is. There is not the same universal tendency to teach. Nevertheless the desire to teach, even the need to teach, can be extremely powerful—think of parents teaching their children to walk or to talk. In other words the need to teach is also innate and universal.

No one can learn for someone else. You cannot magically impart understanding to another. Learning, as I have repeatedly stressed, is intrinsically difficult. The student must face that difficulty on his or her own and cannot delegate it to anyone else not even to the best and most inspiring teacher in the world. On their side the teacher must understand their proper role, which evidently cannot be to learn for the student or to magically remove all impediments to learning. Rather it is to support and encourage the student in the difficult task of learning; to help them face what must be faced; to frame the whole experience so that it is seen as an exhilarating journey rather than a burden; to admit openly to the students that everyone—even a teacher—experiences setbacks and barriers in his own learning experiences.

A good model for teaching was given to us by Plato's account of the Socratic method. It consists of asking questions and eliciting a response. Of course not any old question will do. Questions must be tailored to the student, the subject, and the situation. To work in this way involves a creative and spontaneous response to the immediate situation. At its best and most brilliant the right question at the right time evokes the insight that the teacher is trying to bring forth. Socrates always maintained that his role was that of a midwife. He was just helping the student to become consciously aware of something that she knew implicitly all the time. Whether this foreknowledge is real or not, it accurately suggests the subjective feeling of the creative process and it acknowledges that learning is not the transfer of information but a creative process of insight. It puts the respective roles of the teacher and student into the correct perspective.

Perhaps the best way to think about teaching is as another form of learning—the teacher also learns something through the learning of the student. Recall the different kinds of learning that I discussed in Chapter 5. It might have been more complete if I had added teaching to the list. It is common for teachers of mathematics and other disciplines to report that teaching a subject raised their understanding to a new level. I remember teaching an analysis class as a teaching assistant at the University of California at Berkeley and suddenly realizing that I now understood the concept of continuity and the theorems associated with it in a way that made my previous understanding trivial and superficial. A great way to learn a new subject is to teach a course in the subject.

Of course there are other ways in which teaching can be considered a form of learning. Good teachers must be able to project themselves into the position of the student. In the vocabulary that I have been using, he or she must be able to see the situation through the operative conceptual system of the student. Maybe this has to do with empathy or mirror neurons but it is definitely a skill that is not very common. The reason for this is that a good teacher must be able to function in an ambiguous situation which contains her own understanding on the one hand and the varied understandings of her students, on the other. This is not an easy thing to do. Much more common is for the teacher to feel that her way of looking at things is the "right way," or the "only way," and therefore to be totally unable to imagine how anyone could see things differently. As a result, the student is made to feel inadequate or stupid which only impedes his or her progress.

9.2 Why Have Teachers When You Have the Internet?

There are people today who ask why teachers are even necessary. They say that learning can happen without teachers. If education merely involves the transfer of information, then wouldn't it be more efficient to do this electronically? When it comes to the teaching of calculus, say, why don't we just make a video of the lectures of the best calculus teacher and make it available to everyone over the internet? Many universities are already making many of their courses available online. There are also many interesting computer based courses that not only expose the students to the required material but also guide them through this material and force them to demonstrate a mastery of preliminary stages before they are allowed to proceed to the new step. Such courses have strengths that go beyond what the traditional classroom can offer. For example instruction can be more individualized. It can be tailored to students with different strengths and backgrounds as opposed to traditional lecture courses that tend to be one-size-fits-all affairs. Students often enjoy computerized education because it gives them the freedom to set their own schedule and advance at their own pace. For institutions facing increasingly onerous financial constraints computerizing education may feel like a solution made in heaven.

Administrators and politicians often think that computerized education means better education at a more reasonable price. Will the education system move in this direction in a major way in the future?

Will ten of thousands of educators lose their jobs in the process? There are powerful forces pushing society in this direction. As a consequence, teachers and society in general may be forced to confront the questions that I am discussing in this book. What is the goal and purpose of education? What is learning? What is the role of the teacher and what does her physical presence contribute to learning?

We will never be able to have a thoughtful discussion about the appropriate role of computer-aided education without a simultaneous consideration of the nature and value of teaching. I said that I was interested in "creative learning" and it follows that I must also specify what I mean by "creative teaching." If by education we are referring to the transfer of information, if we mean rote learning or the mastery of standard algorithms, then it seems possible that technology is the way to go. To appreciate the value of having a human being teaching a class we must look on the teacher as a facilitator of "creative learning." When teachers begin to see their main task as fostering the creative development of their students then their vital role in the educational process will become clear. If they see themselves merely as the facilitators of information transfer then machines that do the job more efficiently will someday replace them. Of course good teachers can still make appropriate use of technology. The question at issue is really the value of having the regular and proximate intervention of human teachers as opposed to relegating them to creating software or responding to students online.

Creative teaching consists of nurturing creative learning. It is very much a creative activity in its own right, one that involves being a midwife to the kinds of learning changes that I have discussed: mastering concepts and conceptual systems, and especially, transitioning from one conceptual system to another. One thing is immediately clear. You cannot teach what you yourself have not mastered. You cannot teach concepts that you yourself only grasp weakly. One needs only consider what too often passes for the teaching of mathematics at the decisive elementary school level. There are too many people who teach mathematics at this level with little or no disciplinary training, who themselves suffer from "mathematics anxiety" and freely admit, "I was never very good at mathematics." These people don't really understand what they are teaching or why it is important. As a consequence, what often gets transmitted to students are vague or confused mathematical ideas with an ever present aura of anxiety, projected from teacher to student, that makes the subject seem unpleasant and even threatening. As a consequence mathematics is scary to many intelligent children (and adults, for that matter).

If teachers have little conceptual grasp of the subject they must rely on rote. I can remember years ago one of my children's elementary school teachers telling me how she was using Venn diagrams to explain the properties of sets. Usually the diagram that is used contains three overlapping circles, each representing a different set or collection. I asked what diagram she used when there were only two sets. She was taken aback and maintained that there had to be exactly three circles—not two or four. Clearly what had happened was that she had memorized some "rules" for handling Venn diagrams. What had been introduced into the curriculum as a conceptual aid had degenerated into another arbitrary fact to be learned by rote. This is typical of what happens with the teaching of mathematics and science and the result is the difficulties in student learning that are generally acknowledged to exist. It is not that the subject is intrinsically so difficult; it is that the teachers have not been educated properly and therefore they in their turn fail their students. The solution to these problems will only come with better teaching and teacher training. The kind of teacher training that is required would aim at "creative teaching for creative learning." It would emphasize conceptual development and conceptual change for both teachers and their students.

One wonderful thing about mathematics is that it is common to encounter children who are very bright and curious about mathematics. They are capable of asking good questions; questions that have substantive mathematical content and so are necessarily subtle and difficult. Why is $-2 \times -3 = +6$? Why can't you take the square root of a negative number? What do the dots mean in the expression ***1/3* = *.333…***? Why is π a number? What is chance? One of these seminal questions is the one that has been coming up repeatedly in this book, "What is a number?" All of these questions are wonderful but none of them have easy, pat answers. Each one is an invitation to explore mathematics more deeply but they cannot lead anywhere unless the teacher validates them and the teacher cannot validate them if she has never considered such questions herself.

The very idea that mathematics is something that you can talk about, something that is actually interesting, that there are easy-to-state problems in mathematics that no one knows the answer to, is alien to most classrooms. It is often easier to close down such questions by pleading time constraints or by putting off the question to some later time (which, unfortunately, almost never comes). Students then learn that asking "good" questions, that is questions that are open-ended and are conceptual, is not appreciated and they often respond by closing down. It is terribly sad when this happens. Many bright kids are lost to science in this way.

9.3 Teaching Concepts

Let's go back to a point that has been coming up over and over again, namely, that creative learning is always challenging, and consider the role of the teacher in this context. Now no one can make a person learn and no one can learn for someone else—everyone has to learn for himself or herself. Cognitive changes happen in our own mind and brain—no one as yet can step into our brains and make the requisite changes to our brain circuits. As a consequence, learning must be active; there is no such thing as passive learning. What are the implications for the role of the teacher? One might simply conclude that if the student can only learn for herself, there is no need for the teacher but this response would be unfortunate, simplistic, and misleading. The fact that creative learning is difficult and personal makes the good teacher all the more important. The more difficult the task the more important it is to have a guide—someone who has covered the same terrain in the past and remembers the problems and potential pitfalls, who can say, "I see where you are coming from but why don't you try to think of it in the following way?" The good teacher cannot learn for her students but she can inspire them and encourage them when things get tough.

There is no royal road to learning. How do you go from thinking of a number as a counting number to thinking of a number as a fraction? How do you make that leap? How do you help someone else make that leap? It certainly does not help to say things like, "It's obvious." or "It's logical." The whole problem is that it is *not* obvious or logical to the student. In general understanding does not only depend on the analytic intelligence; you cannot reason your way to conceptual change. The situation that presents itself is that the student has a way of thinking about numbers that has served her well in the past. Then she is presented with a new situation in which this way of thinking does not work and is challenged to acquire a totally new way of thinking about the world of numbers. It is the teacher who presents the challenge, who demands that the student learn to think differently. The process begins with the teacher who sets the stage for learning by making a demand that the student cannot meet at their present level of mathematical development.

I mentioned earlier our erroneous sense that education is exclusively about *gaining* and that we forget that it is equally about *losing* our sense that our current conceptual system is the only way to think about the situation. It's hard to give up your hard-won ways of thinking. The initial role of the teacher is the negative one of breaking down the student's cognitive equilibrium. The mathematics educators David Tall and Eddie Gray[i] have a great example of the difficulties that arise out a student's unwillingness to let go of a familiar and comfortable point of view. This particular student was really good at adding. She was proud of this skill and had been well rewarded for it. When multiplication was introduced as repeated addition she (quite reasonably) multiplied by translating every multiplication problem into a problem in addition. Later, when faced with more complicated problems in multiplication, this approach first became onerous and eventually broke down completely. The problem here was not that the child was stupid—it was that she had become inappropriately attached to addition. The teacher needed to help her see that her old way of thinking was not suitable to the new environment of multiplication and gently help her to learn to see things in a new way—you don't have to go back to addition to solve every multiplication problem.

By becoming aware of the demands that are implicitly being made of the student, the teacher is in a position to appreciate the struggles that some of them will experience. The "best" students are not necessarily those that grasp the new procedures most readily and efficiently. The best student may well have serious trouble with the new concept because she sees the problems and potential pitfalls most clearly. After all why *is* *2/3* a number? Why *is* $-2 \times -3 = +6$? These are mathematically interesting questions. It takes a really bright child to be troubled by the difficulty here.

We will only appreciate the role of the teacher if we consider our own response to problematic situations. Suppose I have to do some small but unfamiliar repair around the house—something I am not so good at. I try it but inevitably something comes up that I had not foreseen. If I cannot solve the problem right away I often become irritated—just not knowing how to do it right bugs me and, to break the tension, I may do something impulsive and make things worse. I spoke earlier about how it can feel humiliating to be placed in a new learning environment. Irritation, humiliation, and tension are all normal responses to finding oneself in the kind of learning environment where you are forced to admit, if only to yourself, a certain inadequacy. Learning, at its most significant, involves a kind of cognitive dissonance, almost a double bind. "A number is (can only be) a counting number but the teacher talks as though ***2/3*** is a number. How can this be?"

The teacher's first role is to shake up the rigidity of the student's present conceptual system and in this way make room for the new. She must take conscious responsibility for what she is doing so as to be able to anticipate and correctly evaluate the kind of reactions that are normal when one is pushed into such situations. Who teaches students how to deal with the stress that arises? Good teaching involves a great deal of stress management. It is necessary to strike a delicate balance. Too little stress and everyone goes to sleep; too much stress and the students will stop engaging with the material. Because every individual has his own optimal level of stress, the good teacher has to be a kind of virtuoso in her management of the class. The teacher's difficulty is compounded by the fact that students, like all people, are usually uncomfortable with stress, and so the teacher who does not challenge the students may be more popular, at least initially, than the one who does.

Confronting some new concept that one does not understand is often accompanied by self-doubt and other negative feelings, "Maybe I'm not smart enough to get this. I was never very good at mathematics." Who helps students understand that these so-called negative reactions are a normal and inevitable part of the learning process? Who tells the students that such reactions are actually "good" because they indicate an authentic engagement with the challenge of learning and that no "creative learning" can ever occur without engaging in this way? Who helps them develop the "stress tolerance" that, more than IQ, will go a long way towards determining who will become a good learner and a creative person?

Only a good teacher can do so. Such teachers can help a student frame her reactions constructively or reframe her negative reactions. Remember that the very same tension can be experienced either as an unpleasant stress to escape from or as an exhilarating excitement to enjoy—we *learn* to experience tension in one way or the other. A good teacher, who remembers what it is like to encounter such situations (and hopefully is still encountering them), can help students to see these situations positively and constructively and learn to enjoy them. This is something that a program will never do—look into the faces of students and empathize with what they are going through.

On the other hand there is the poor teacher for whom the entire subject is "obvious" and "objective" and can only be understood in one way—their way. The poor teacher is unaware of or unsympathetic to what the student is experiencing. In good teaching the cognitive difficulties of the students is matched by the cognitive difficulties of the teacher authentically confronting the problem of how to teach. Good teaching is an ongoing and unending creative activity—each class is different; each student is unique and represents his or her own challenge. It is never a simple question of right or wrong, intelligent or stupid. All students have potential and their own unique internalization of mathematics. How do we reach them and help them to modify their internal structures?

9.4 Teaching Hierarchical Conceptual Systems

I've been looking at mathematics as a series of conceptual systems. You cannot think about learning and teaching mathematics without focusing on the conceptual. We have seen that it is not enough to study such systems in isolation from one another but we must also look at the relationship between various conceptual systems, especially how a more advanced system, CS_2, develops out of the more basic system, CS_1. Our prototype is this regard was the counting numbers and the fractions or the fractions and the real numbers. CS_1 and CS_2 are tied together in a hierarchical relationship and the complexities that arise in the teaching of such transitions are the subject of this section.

When I discussed the difficulties inherent in teaching and learning CS_2 I was referring to an incompatibility between two conceptual systems. Because the teacher is firmly based in CS_2 this incompatibility is often invisible to her. Yet without an appreciation of the nature of the difficulties that the student experiences—that they are based on seeing this piece of the mathematical world through the lens of CS_1, a point of view which is different but nevertheless is relatively consistent and complete in its own terms—there can be no good teaching.

However CS_1 and CS_2 are not totally isolated from one another. In terms of mathematical content CS_2 can be derived formally from CS_1. For example, given the counting numbers as CS_1, then the fractions can be considered as ordered pairs (n, m) of counting numbers, where (n, m) is equivalent to (p, q) if and only if $nq = mp$. In these terms a fraction would be an equivalence class of ordered pairs. (The equivalence class for one-half would include not only $(1, 2)$ but also $(2, 4)$, $(3, 6)$, $(4, 8)$, and so on.) Then we can define addition and multiplication of pairs so that the resulting system is basically identical to the fractions. Of course this is a formal and logical derivation. Even if you follow the logic this does not mean that you have a conceptual knowledge of "fraction." But this kind of procedure is normal in mathematics. In the same way, we can start with the fractions as CS_1 and define a real number as a (converging) infinite sequence of fractions.

However we are not interested in the formal derivation of one system from another so much as we are interested in their relationship as conceptual systems. Many of the conceptual systems that I have discussed answer the question, "What is a number?" Therefore the terrain that CS_1 covers overlaps with that of CS_2. CS_2 contains within itself an isomorphic representation of CS_1 where the word "isomorphic" is being used in a way that is inspired by the mathematical term but not identical to it. When it comes to conceptual systems even when a student masters CS_2, CS_1 is not eradicated but continues to operate. If we are familiar with the system of positive and negative integers we still, on occasion, think about 2×3 as an area or as repeated addition. In the example, ***.999... = 1***, that I spoke of earlier, one of the problems is that the infinite decimal triggers the conceptual system of the real numbers while the '***1***' triggers the counting numbers. To compare them it is necessary to see that they both can be seen to exist within the same conceptual universe. At a more basic level, I noted in Chapter 1 that our core mathematical systems continue to influence us throughout our lives.

One of the reasons mathematics is so complex is that the same mathematical object—the integer ***2***, say—can be thought of as belonging to multiple conceptual systems: counting numbers, integers, fractions, real numbers, complex numbers, and so on. We have a sequence of conceptual systems CS_1, CS_2, CS_3, CS_4, and so on where every object in CS_1 also lives in CS_2, CS_3 and CS_4. Furthermore the earlier systems coexist with the latter ones so that we are free to refer back to representations at the CS_1 level even when we are ostensibly operating within CS_4. Another way of saying this is that every mathematical object has multiple representations—this is the basic ambiguity of mathematics. It is not sufficient to discuss objects in isolation; one must consider them within an appropriate mathematical context or rather within various contexts at the same time. This is why mathematics courses are inevitably hierarchical—you keep going back and reconsidering old ideas from a "higher" point of view.

One way that hierarchical structures are established is through the process of abstraction in which properties of one or more 'concrete' situations, CS_1, are generalized to produce a new point of view, CS_2, (as I discussed in Chapter 3). For example number systems like the fractions or the real numbers are generalized to "mathematical fields." Or one focuses on the distance between real numbers and then generalizes the real numbers to "metric spaces," which are spaces equipped with a generalized idea of distance. When one moves from a concrete base system to a more general abstract system in this way, the old system, CS_1, is often thought of as a mere "example" of the generalized structure CS_2. But cognitively it does not work that way because the old system keeps functioning within the new situation. In practice the new and the old are tied together and one keeps going backwards and forwards between the two. If one tries to understand something within a abstract CS_2 (like a field or a metric space) or attempts to determine whether some hypothesis in CS_2 is valid one goes back and checks it out in the various CS_1 systems which CS_2 generalizes. The power of CS_2 is that it is one system that incorporates a whole series of CS_1 subsystems.

At this stage we have uncovered another reason why teaching and learning are so complex. From one point of view CS_1 and CS_2 are incompatible with one another. A child in CS_1 will give you a different answer to a question than a child who is in CS_2 (and both answers will be "right"). However the incompatibility does not lie in the same horizontal plane as would be the case if you were describing something in two different languages. The systems are hierarchical, that is, CS_1 can be represented within CS_2. Constructing CS_2 includes creating a compatibility with CS_1. Moreover the view from CS_1, so to speak, keeps on functioning even within CS_2. If someone asks you to multiply two numbers in your head you go back to the old conceptual universe of the counting numbers in just the same way that you use Newtonian mechanics and not the theory of relativity if you are sending a rocket to the moon. The old systems stick around and are available for use in the appropriate situation. The different representations are also available but cease being a problem; $.999\ldots = 1$, for example.

What are the implications for teaching such hierarchical systems? In the first place the teacher must appreciate that: (1) CS_1 and CS_2 are logically compatible because an isomorphic image of CS_1 can be embedded in CS_2 (the counting number '***n***' is the integer '***+n***') but (2) CS_1 and CS_2 are conceptually incompatible because ***2*** times ***3*** 'is' an area whereas ***−2*** times ***−3*** is not, that is, multiplication of integers is a different conceptually than multiplication of counting numbers.

This incompatibility remains with the student for a long time. You can never dispense with the student's old way of thinking. The teacher must aim not only at introducing the new concept but also at integrating the new with the old. Otherwise the two systems will operate independently of one another and the result will be unnecessary cognitive conflict. All too often the process of integration is omitted and the teaching of each new course or concept starts from scratch. This is one of the consequences of the formalist philosophy of mathematics. It is the idea that you can always develop a mathematical subject from a set of axioms, from zero so to speak. Cognitively this is never possible; there is no zero stage. Teaching involves remaining aware of the conceptual foundations that every student *must* inevitably bring to the table. In every situation the teacher must take the time to investigate what the student means by number, by function, by continuity, and so forth. This is hard to do. It is not enough to lecture at your students—you must ask them what they think and attempt to discern what they understand. We are not teaching objective material, as so many believe; so much as we are helping students to develop appropriate conceptual systems.

9.5 Continuous or Discontinuous Teaching

In my discussion of learning, I emphasized that conceptual learning contains both continuous and discontinuous elements. There is room both for rote learning and sudden bursts of insight. The continuous element involves mastering computations within a given conceptual system—you understand the counting numbers and the addition of small integers and then you learn the algorithm for addition. Such computations increase one's familiarity with, and intuition for, the conceptual system in question. The rules of classical logic apply and, in particular, this is the context within which one emphasizes precision. It is at this level that things are either right or wrong. Systematic errors are important because they imply that the student's internal conceptual system is not a good fit for the one to which the community subscribes.

The discontinuous element involves going from one system to another. The rules of classical logic do not apply, that is, you expect the student to find things problematic; you anticipate contradiction and even paradox. Such situations are always ambiguous because there are two inconsistent views of the same mathematical terrain floating around.

These two very different kinds of learning inevitably require different kinds of teaching. Most of the teaching that I have been discussing in this chapter involves situations of discontinuity. On the other hand most of what goes on in the classroom and lecture hall involves continuous teaching and learning. Most of the teaching of mathematics at all levels—elementary, secondary, and undergraduate—is of this type. So is the teaching of mathematics to students in other disciplines—statistics to students in psychology, differential equations to future engineers, calculus to students in economics and commerce. Unfortunately it also includes most teaching of mathematics to mathematics majors.

It is here that we see the consequences of incorrectly conceptualizing the tasks of learning and teaching. From our perspective as "experts," endowed with what we think of as the "correct" or "objective" way of thinking about the subject, we forget that learning is a process of repeatedly replacing CS_1 by CS_2. The good teacher must constantly remain vigilantly aware of the dynamic nature of learning—anticipate it, expect it, and encourage it. If it is ignored then one might as well throw in the towel and admit that there is no real education going on.

9.6 Mirror Neurons: Teaching and Learning as Social Activities

One of the most intriguing discoveries in modern neuroscience has been mirror neurons—circuits in our brains that reflect the circuits in the brains of others. We seem to be connected to other people in a very intimate and direct way. We naturally feel what others feel and think what they think. A kind of resonance between ourselves and others is involved. The phenomenon of mirror neurons is the biological correlate of the fact that human beings are social animals. There is much speculation that large brains evolved as a result of the complexities of social interaction. This accounts for a basic duality that is at the heart of human consciousness. We are simultaneously intensely individualistic—

separate and distinct from others—but, at the same time, we are deeply enmeshed in multiple social groups—from families to nations to the entire global society. Learning and teaching are best viewed from this perspective. It is both individual—something that happens to you in your head—and social—something that only has meaning in a social and cultural context. Good teaching and learning must address both of these perspectives.

One consequence of the social dimension of learning is that the presence of a human teacher is essential. To a considerable extent we all learn through imitation. We understand things in the way our parents, teachers or professors understand them. It is a cliché to say that most people teach in the same way that they have been taught. This can be bad news if the teacher is inadequate but, on the other hand, it can be very good news indeed. The more we interact with our teacher the more we are privy to her thought processes, intuition, and conceptual systems and so the more open we will be to what she understands, how and what she thinks. The student is not merely internalizing the information that the teacher provides but is being directly affected by such things as the teacher's body language, her enthusiasm for and love of her subject. The teacher cannot fake it. Nonverbal cues count as much as what is made explicit; what she doesn't say is as important as what she does say. If she is fearful or anxious then she communicates this fear or anxiety. If her understanding is superficial then superficiality is communicated. If she has a deep understanding of the subject then she will communicate her confidence and enjoyment.

To be a good teacher requires courage and openness. There is nowhere to hide when you stand in front of a class—what you are and what you know are visible to everyone, especially yourself. Does the teacher have the confidence to admit ignorance and fallibility and answer a question by, "I don't know the answer to that but I'll think about it and get back to you next time." Teaching is a profession with great rewards in terms of job satisfaction but there are major demands to be met. If the learning situation requires the students to allow themselves to be vulnerable in order to open themselves up to new approaches and perspectives then it also makes the same demand for openness and vulnerability on the part of the teacher. There will always be the temptation to hide behind authority or the formal aspects of the subject, in other words, to close down. There will always be the temptation to be intellectually lazy—to plod through the same material class after class and year after year—and ignore the particular needs of the group of students in this particular class. Every class and every student is a different challenge. The most rewarding thing in the world is to see the light go on in the eyes of the student and hear them say, "Ah, yes. I get it!" Ultimately teacher and student are on the same human level. The student is continually learning how to learn and the teacher is constantly learning how to teach, both of which come down to the same creative process.

A good teacher is a rare commodity, extremely rare. Such people are as important as good scientists and good artists—they *are* scientists and artists in their own right. How can we attract the right kind of people into teaching? How can we train them so that they understand what is at stake? How can we support their efforts to teach creatively, to create and maintain that very tricky and subtle classroom atmosphere that is required for real learning to go on? Can we accord this profession the respect that it deserves and restore it to a central position in society? In a society that ostensibly worships innovation let us remember that an educational system that is worthy of its name would foster creativity at every level from elementary school on up. Creative learning would then be seen as the lifetime activity that it must be in the modern world. Let us never forget that the society of the future is being created student-by-student by the teachers of today.

---

[i] Gray and Tall (1994).

Chapter 10

A Note on Artificial Intelligence

10.1 Introduction: Why Discuss AI in This Book?

I had long intended to write an article about the implications of artificial intelligence because I had a feeling that the perspective from which I've been writing about mathematics and science is relevant to a discussion of AI. In general, when I sit down to write something, I don't begin with a predetermined position and put it down on paper. Often I write in order to clarify what I think for myself. The process has been successful if, at some point, I can say to myself, "Ah! So that's what I think." (The discussion of creative intelligence having its origins in the non-conscious explains why this is so.) This attitude, I have found, leads to the best kind of writing, and this is what is going on in this chapter. I have the sense that there is something that I have to say about AI, but for the moment, I am not sure what it is.

Any writing about AI is necessarily tentative because things are in flux. We are at the beginning of a technological revolution in which startling developments are appearing, and no doubt, will continue to appear at an accelerated rate. No one can say anything definitive about where it will end or even about the process of getting there. However everyone seems to agree that everything, absolutely everything, will change.

Why does a chapter on AI have a place in this book? What is the connection between AI and the rest of the book? This wasn't clear to me at the outset but the connections have now become clearer and almost obvious. The theme of the book is creative intelligence, its properties and characteristics, and the way it manifests itself in the process of deep thinking in mathematics and science. Inevitably then, the book must contain a reflection on the nature of intelligence and creativity. The existence of artificial intelligence makes the meaning of "intelligence" a matter of great urgency. When we use the word "intelligence" in referring to a computer program, a mathematical proof, or an AI system like ChatGPT, is the "intelligence" that we are referring to identical in all cases? I think not.

Can an AI system possess the kind of creative intelligence that is discussed in this book? If the answer to this question is "yes, they are essentially the same" or even if the answer is "yes, they will one day be the same" then we will soon be living in a world where intelligent machines can (or will be able to) do creative work in math and science. But even if the answer to this question is "no, they are fundamentally different" then there remains the interesting question of the relationship between the two, that is, between artificial intelligence and the creative intelligence that appears in the examples of deep thinking that I have been describing. That there is a connection is clear, if only in the way the expressions "intelligence" and "deep thinking" are used in both contexts.

The importance of these questions is highlighted by the original mission statement of the DeepMind group at Google: "Solve intelligence and use it to solve everything else." Thus, in the AI world, the question of intelligence is primary. However the Google researchers do not specify what they mean by intelligence, nor what it would mean to "solve" it. What is intelligence? Is it even a technical problem that can be solved? Behind the work of such AI research teams there must exist a definition of intelligence. What is it? Is it identical to creative intelligence? These are some of the questions that come to mind.

10.2 Creativity and Artificial Intelligence

I sense that the nature of creativity and the question of whether or not an AI system is capable of creativity is always implicit in discussions about AI. We know that ChatGPT is intelligent in certain ways, but is it creative? That is a fundamental question, and it is difficult because, as I said earlier, there is no consensus on the meaning of either intelligence or creativity. Of course, it is the purpose of a book such as this to make the meaning of these terms a little clearer or at least to make people aware of the subtle implications of these ideas.

Because creativity is key in many ways to this discussion, I shall take it up now. Douglas Hofstadter, the author of the amazing book *Gödel, Escher, Bach*, who is acknowledged to be the father (or grandfather) of AI, made the following comments about the 2014 state of AI, as reported by his colleague Melanie Mitchell in her fine book *Artificial Intelligence*. He said, "I am terrified. Terrified." (These comments were made at a conference on AI that took place at Google in 2014.) What terrified him? He gave an example of a music program by the musician David Cope that could create music in the style of a famous composer, Chopin, for example. For Hofstadter, the music that he loved felt like "a direct message from the emotional heart of the human being who created it. … The idea that pattern manipulation of the most superficial sort can yield things that sound as though they are coming from a human being's heart is very, very troubling." He told a story about asking an audience at the prestigious Eastman School of Music which of two pieces a pianist played was a piece composed by Chopin and which had been composed by a computer program. The audience, which included various faculty members, guessed wrong to the consternation of all concerned. This in itself should tell us something about the potential impact of AI, for it anticipated the phenomenon of "fake news" which is now so prevalent that people have great trouble sorting what is real from what is not.

Hofstadter said to the group at Google, "I was terrified by WMI (the computer program), terrified. I hated it, and was extremely threatened by it. It was threatening to destroy what I most cherished about humanity." Remember, all of this was happening in 2014, which is the ancient past when it comes to developments in AI. But the general response is well worth taking seriously since it has been repeated with great urgency by some of the founders of the AI movement.

This then is the situation that confronts us with respect to the AI revolution. Are we about to lose the essence of our humanity? Is there no difference between the intelligence of an AI system and our own human intelligence? Can the artificial creativity of an AI system be distinguished from the creative intelligence of Einstein, Newton, Bach, and Beethoven? These are questions that are non-trivial, as we would say in mathematics.

Hofstadter is not the only expert who is concerned. Many others have gone on the record with their concerns and petitioned the U.S. Congress to consider legal regulation of the industry. Because the claims and the ambitions of the AI community are so wide-reaching and open-ended, people are right to be concerned. If "solving the problem of intelligence" means producing a self-conscious, autonomous machine intelligence that is capable of developing further on its own in a kind of artificial evolution, then we are indeed in a radically new world populated by a new form of sentient being. Frankly, I don't think that this scenario will ever be realized, but whether or not it actually happens tomorrow or a hundred years from now, the time is assuredly ripe for us to think about it seriously and not just react after the genie is out of the bottle as we did in the case of the development of atomic power or other scientific breakthroughs. This chapter is my contribution to this discussion.

10.3 What is a Human Being?

There can be no more fundamental question that this one, "What is a human being?" One of the defining qualities of human beings is surely self-consciousness. Human beings are not only conscious but self-conscious; they have the ability to be conscious of themselves, that is, to stand outside of themselves, so to speak, and see themselves as an object in the world. With the advent of language, which is associated with self-consciousness and has surely led to its deepening, we have invented the word "I" to represent our own self-consciousness to ourselves. To ourselves we are the "I," but to others we are represented by our name. This is so important that we spend a great deal of time and effort inducing our children to identify with their name. One might argue that this ability that human beings have to objectify themselves in this way is our defining characteristic. It has many consequences, good and bad. The good includes respect for the individual and the idea that all people are equal before the law. The bad or unfortunate side effect of this objectification is bifurcating people into two parts, which we could think of as our subjectivity, on the one hand, and its objectification, on the other. This inner split can be experienced as an inner rupture which, as a consequence, produces the yearning to be made whole again. This whole drama is, of course, the subject of the Garden of Eden Myth, which, in some ways is the defining mythology of Western civilization. In this view, it is self-consciousness that is the cause of humanity's loss of the Garden, which represents our primordial unity with the world.

There is a very powerful tendency in the modern world which maintains that self-consciousness is an epiphenomenon of neural complexity. It would follow, as some AI visionaries claim, that if and when we pass Ray Kurzweil's technological singularity, we will share the planet with intelligent machines. However, there is another possibility, a more traditional point of view (which I am trying to clothe in modern language), that would say that intelligence, at least the creative intelligence that is the subject of this book, is not the outcome of complexity at all. This view would see intelligence as built-in, so to speak. It would reverse the AI view and see complexity as the result of an intelligent and creative process. Then intelligence would be a process that produces complexity.

One of the dangers we face in the modern world is that we will come to allow the activity and output of AI systems to define human consciousness. This is not so far-fetched. There are already philosophers and psychologists who claim rule-based rational thinking to be the essence of mind, to be the way that the brain functions. If such thinking becomes ever more dominant, then what I call creative intelligence or deep thinking might disappear from view, that is, from our collective consciousness, because non-conscious elements, such as the ones that have a role in creativity, would cease to be visible. This would happen in the manner that is normal when one gives something a name. Naming something highlights the object, and as a result, we become conscious of it, whereas things that have no names sink into the background and into what I was calling the pre- or non-conscious. This is the notorious "I was looking right at it but I didn't see it" phenomenon. My fear is that creative intelligence will sink into a non-conscious background, and people will forget that such a thing exists. This would constitute a great loss for us all.

However, such abstract, philosophical matters, as important as they are, have been overtaken by questions that carry immediate and very practical implications. The very foundations of our society are in question, not to speak of the livelihood of millions. The foundations I am speaking of that are in question include the notion of personal freedom, including freedom of speech and religion. In fact, our society's continued existence as a liberal democracy is under direct attack as I write these words. The new technological revolution will have a direct impact on each of these things, and the future is unclear, to say the least.

The theme of this book is that the essence of human nature is directly connected to creativity and intelligence. In fact, you could say that self-consciousness is by its very nature creative. So the connection between my description of creative intelligence and this new variety of *artificial* intelligence is interesting and timely. There is clearly a connection between human intelligence and this new machine intelligence since, for many observers, like Harari, an AI system is clearly intelligent. However, connected does not mean identical.

I must admit that in an earlier version of this material, it appeared obvious to me (as it once seemed obvious to Hofstadter) that computer-based systems were not intelligent and certainly not creative. I now realize that things are not so simple. In the light of experiences like Hofstadter's and my own experience of ChatGPT that I'll discuss below, I have revised my initial reaction and made it more nuanced. If anything, this makes the questions raised even more important for our collective future. The AI revolution is coming. There can be no turning back. Thus, it is imperative that we collectively distinguish the possible benefits from possible dangers. And it all comes down, in my view, to the question, "What is intelligence?" This is one of those questions that seem obvious at first glance but become more complex and opaque the more you ponder them.

10.4 ChatGPT's Take on What I Think about AI

My niece, the psychologist Krista Heinlein-Byers, was visiting a while ago and the conversation turned to AI. She didn't try to explain AI to me but immediately opened ChatGPT on her phone and asked it to write a hypothetical essay about AI in the style and with the ideas of William Byers, that is, me. The following is what came out (unedited or changed in any way):

The Question of Artificial Intelligence

In recent years, artificial intelligence has captured the imagination of the public and the intellectual community alike, heralded as the next frontier in human progress. The promises made in its name are grand: machines that think, that reason, that may one day surpass human intelligence. But these claims, as with many technological marvels, warrant a deeper investigation—not merely into what AI can do, but into what it cannot.

To understand artificial intelligence, we must first reflect on the nature of intelligence itself. Intelligence, as it is commonly conceived, has been reduced to an ability to solve problems, to compute, to analyze data in increasingly efficient ways. This, undoubtedly, is something AI excels at. It processes information at speeds unimaginable to the human brain, finding patterns in chaos that elude even our most talented minds. And yet, this very efficiency is its limitation.

Human intelligence, in contrast, thrives on ambiguity. It is not merely a computational act but an experiential one, deeply entangled with the way we live and perceive the world. Our ability to navigate uncertainty—indeed, our ability to remain within it—defines much of what it means to think as humans. We are not machines of perfect logic. Our decisions, our breakthroughs, our deepest understandings emerge from a space where reason meets intuition, where clarity wrestles with confusion, and where contradictions coexist.

Artificial intelligence, by design, lacks this engagement with ambiguity. It is built on algorithms that function within well-defined parameters, excelling in environments where the rules are clear, the data ample, and the outcomes measurable. Yet intelligence—true intelligence—is born in precisely the opposite conditions. It emerges where there are no clear rules, where the data is incomplete, where outcomes are uncertain. This is the territory of creativity, of invention, and, ultimately, of human flourishing.

One might argue that AI is not meant to replace human intelligence but to augment it. This is a reasonable perspective, and indeed, AI holds tremendous potential as a tool. However, we must be cautious in our enthusiasm. If we come to equate intelligence with the ability to compute solutions, we risk losing sight of what makes human intelligence so rich, so nuanced. We risk forgetting that intelligence is not simply the capacity to arrive at answers but the capacity to live with questions.

AI, for all its impressive capabilities, operates in a world that is bounded. It is a world of black and white, of if-then statements, of inputs and outputs. But the world we inhabit is one of shades and gradients, where meaning is rarely fixed and certainty is the exception, not the rule. In this world, intelligence is not about eliminating ambiguity but about embracing it. It is about seeing the beauty in uncertainty and understanding that the unknown is not a problem to be solved but a space to be explored.

As we continue to develop AI, we would do well to remember that the most profound insights—those that reshape our understanding of the world—do not come from a place of certainty. They come from the edges of our knowledge, where clarity fades into mystery and where questions, not answers, reign supreme. If we forget this, we risk not only misjudging the nature of intelligence but impoverishing the very experience of being human.

-ChatGPT Describing What William Byers Thinks about AI

When I read this, my first reaction was, "I agree with this. It's good!" It's an excellent synthesis of what I think and have written about in my books. For example, that true creativity has its roots in the problematic, not in rule-based systematic thinking. Thus the need to emphasize process as well as content, questions and not just answers, mystery and wonder and not only facts. AI systems are surprisingly good and will get better. But their strength may tempt us to settle for an impoverished version of intelligence and creativity. I'll expand on this in what follows.

I will attempt to clarify matters by calling ChatGPT's take on my thinking, "Bill Byers' ideas," whereas when I use my own words and ideas (at the time I am writing this) I will call them "William Byers' ideas." When I say "I," the reference is always to William Byers. Bill Byers has no "I."

As I said, Bill Byers has interesting ideas. It's always stimulating to read a review of something you have written. It can be difficult if the review feels negative or wonderful if the reviewer thinks you have something constructive to say. But either way you respond to someone else's opinion. Here there is really no opinion, just a synopsis, but I responded in a similar way. However the realization that there was no person on the other end of the line made it feel a little weird. It is a strange feeling to read about your own thoughts and ideas when they are returned to you having passed through the filter of a computer system. I reacted as though there was a someone there, but maybe it is not a person I feel but an intelligence? Is that intelligence that I sense a mere projection of my own intelligence, or is it another, different sensibility? When you read a draft of something you have written, you respond in much the same way. You see your thought from the outside so to speak, whereas when you were writing it came from the inside. This is the normal subjective/objective split, which is implicit in consciousness itself (as I go into a bit later). Maybe what I like about Bill Byers' ideas was similar to my reaction to reading my own thought?

Nevertheless I agreed with much of what Bill Byers had to say about AI. The reader will note that the meaning of intelligence has a central role to play in the discussion of AI. Anyone who has read this book so far will note the extent to which it emphasizes intelligence. Thus, the shift in the title from "Deep Thinking" to "Creative Intelligence" in the thought of William Byers.

Bill Byers stressed that creativity often has its origins in situations that are murky, ill-defined, ambiguous, and uncertain. In my earlier books I gave a lot of thought to ambiguity, constructive ambiguity, that was present in situations where there were two distinct ways of understanding a given situation, two points of view of the same terrain, which contradicted one another. To ask which point of view is correct was, in the situations of mathematical or scientific creativity, the wrong question. To dwell in the contradiction until there dawns a new way of seeing the situation which reconciles the original opposition; this is the creative way of approaching the situation. In the chapter on creativity you will see that many people have pointed to such seemingly contradictory situations, and our approach to them, as the situation which gives birth to creativity. Ambiguity and tension are the key. And, by the way, the key ambiguity present here is the one between Bill Byers and William Byers. Which is the real me?

Intelligence, as Bill and William Byers agree, does not proceed from clarity to greater clarity as would be the case in a mathematical proof. Creativity is not logical! Creativity is a mixture of reason and intuition. As I mentioned in the chapter on creativity, a creative insight involves revelation as well as reason. This motivated me to redefine reason as a creative process and not merely a rule-based one. In this way and more Bill Byers has it right.

For the first time I began to appreciate some of the implications of AI in general, and this particular application in particular. What it does, it does well. It is especially impressive because it is not a synthesis of what I have written about AI since I have not written anything before these words. It is an extrapolation from my writing, *what* I would say if I were to write about AI. Yet as a concise statement of my ideas and voice, it does remarkably well. It is well-written, grammatical, and coherent.

10.5 The Creativity of AI

So far we have seen how an AI system can produce a synopsis, extrapolate out a point of view, and apply that viewpoint to a new situation. Is the AI system doing creative work? Well it feels pretty close. Yet in this section I want to bring the reader's attention to another recent result which I find even more startling.

I discovered recently an article in an internet science journal, *TRENDING*, with the title "*Scientists discover simpler way to achieve Einstein's 'spooky action at a distance' thanks to AI breakthrough—bringing quantum internet closer to reality*," about a study published on December 2, 2024 in the journal *Physical Review Letters*. The authors were studying quantum entanglement using an AI tool PyTheus, that had been specifically created for designing quantum-optic experiments. "The authors were able to train a neural network on a set of complex data that describes how you set up this kind of experiment in many different conditions, and the network actually learned the physics behind it," said Sofia Vallecorsa who works at CERN and was not involved in the research.

The fascinating thing about this research is that "the AI tool proposed a new idea, namely, that entanglement could emerge from a much simpler set of conditions than the ones the researchers normally used." The reviewer said, "Although the scientist were initially skeptical of the result, the tool kept returning to the same solution, so they tested the new theory." And it worked! Isn't this interesting? Here the AI system not only was able to synthesize known theory and results but could suggest a new, improved approach. For most people, this would constitute a creative act. Thus in some sense, AI systems are not only intelligent (the synthesis) but also creative. These kind of situations make a reconsideration of the nature of intelligence and creativity even more urgent.

10.6 What is Intelligence?

Let us go back to the question of intelligence and to the tentative conclusion that an AI system may be intelligent. Let us return to the primordial question of the nature of intelligence. The first thing to note is that there are different kinds of intelligence. The Psychologist Howard Gardner wrote a book called, *Frames of Mind*, about this point. For example there is mathematical intelligence, musical intelligence, spatial intelligence, manual intelligence, and many other varieties of intelligence. When Elon Musk and other high tech entrepreneurs say that they only hire the most intelligent people they are referring to one particular variety of intelligence that they value above all others, namely, analytic intelligence, which seems to correlate quite well with high IQ. However some other forms of intelligence might not require very much analytical intelligence at all. If analytic intelligence implies the capacity to reason, then we are thrown back to reflecting on the nature of reason, which is a complex question that I discussed in Chapter 7.

On the other hand, if intelligence implies the capacity to solve problems then we share intelligence with chimpanzees and even with crows and blue jays. A person can be a mathematical savant with very impressive skills without much analytic intelligence at all. I remember the math faculty members at the University of California at Berkeley when I was a graduate student. Many were brilliant mathematicians but only a few met the standard of genius, which in this case you would define as having made some groundbreaking discovery that had radically changed his or her field. Thus a superior analytic intelligence is not the defining characteristic of genius. Genius requires creativity which brings us back to distinguishing "creative intelligence" from "analytic intelligence."

You can see that coming to a consensus about the meaning of intelligence is not easy to do since it appears that intelligence means different things to different people. In writing this book I was on a search for the most basic kind of intelligence that I could come up with. I am calling it creative intelligence and in a way this entire book consists of my attempt to get a handle on this illusive notion.

In my writing on mathematics I have singled out a form of creative intelligence that I called deep thinking. I originally had in mind people like my Ph.D. advisor, the Fields Medal Winner Steven Smale. His career involved creating entire new fields of mathematics as well as bringing an entirely new perspective to fields that people believed were well understood. In other words I began with the idea that the kind of intelligence in deep thinking came from creative geniuses. As I investigated this form of intelligence I became aware that not only was it characteristic of genius but it was also a capacity that was shared by everyone to some degree. Every child has the capacity for learning very subtle concepts like language and number and the conceptual systems in which they are embedded. A capacity that is so universally shared leads to startling conclusions, namely, that we are all intelligent. That is to say we all share this basic capacity for intelligence. Furthermore we don't have to learn to be intelligent. *We are intelligent!* Intelligence comes with the territory you could say. The forms may very but the potential is universal.

This insight puts the discussion about AI into a very different light. When we ask whether ChatGPT is intelligent, we imagine that we are discussing an independent entity. We divorce the discussion from people, that is, we abstract it from the human condition. The process of abstraction is familiar in mathematics as I have discussed in an earlier chapter. We take some mathematical situation, addition for example, tease out its essential properties, and imagine a structure whose defining characteristics are these abstract principles which we then proceed to study. The great technological innovations that we witnessed in recent years have also involved abstracting out certain human capacities and simulating them with a mechanical device. Society has done this over and over again with great success and these successes have transformed the world.

The new frontier for this process of abstraction and reification is intelligence. However intelligence is not in the same category as speech or vision. Intelligence is too basic and too general. The intelligence being simulated is the intelligence doing the simulation! The entire situation is self-referential, but perhaps that is why these developments are so exciting.

10.7 Is the Output of an AI System Objectively True?

Scientific and mathematical theories appear to be objective. They are written down in a permanent record and seem to point to a reality that goes beyond the mentation of any single individual. In the case of science they seem to imply a permanent, universally valid aspect of the natural world. Take, for example, the theory of evolution. It is a dynamic theory that explains change. Nevertheless, we think of the theory itself as permanent, at least at any given time. Even though we accept that scientific theories change, they have a timeless aspect about them during the period that they remain in favor.

Elsewhere in this book I have discussed the phenomenon of objectivity. To briefly review that discussion, I claimed that there were two distinct kinds of objectivity. Weak objectivity meant independence from things like race, gender, and other idiosyncratic phenomena. Strong objectivity meant that the theory was independent of mind, in the present discussion we could say independent of creative intelligence. I maintained that mathematics and science were weakly objective but not strongly objective.

The question here is whether the output of an AI system, like ChatGPT, can claim to be objectively true. In the first place, it is not clear that such a system is even weakly objective because its results are dependent on its training material. This has been pointed out by many commentators who noted that, for example, a generic "man" for such system would be a white man and not a black or brown man. Of course such anomalies could be corrected if the people in charge were made aware of them and they wanted to spend the time and money to do so. Nevertheless it is clear that any AI system will always contain a certain bias. But let's assume that the good will is there and the money is spent to remove the biases that come to people's attention. Nevertheless bias and error can never be completely removed from such a system, based as it is on databases and, more importantly, as it emanates from a group of people working for a given company mostly interested in making money and increasing stock value.

The next question is weather or not an AI system is strongly objective, that is, autonomous in some sense. When you get the results for some query, you certainly get the feeling that some sort of autonomous intelligence is at work. One is even tempted to anthropomorphize the system because it feels like you are talking to *someone*. Even though it is clear that the program is created by people and that it is trained by people, is the end result, the finished AI system, autonomous? Does this simulation of mind stand alone? And even if it does not stand alone at the present time, does it have the potential to do so at some time in the future? This is a variation on what I have called "the dream of reason," the dream that scientific and mathematical theories stand on their own as absolute truth in such way that they would remain true if all human beings would someday disappear. In this incarnation of this old dream about the stand-alone nature of the products of human intelligence, it is not about the potential of analytic intelligence but the potential of this simulation of the human brain and its potential to become an artificial but autonomous brain in its own right. And it seems to be making progress. Yet we need to remind ourselves that the realm in which an AI system functions concerns itself with only a small fraction of the human situation. AI systems are based on huge data collections requiring arrays of supercomputers using large amounts of electricity. Is the human being a mere collection of data? More likely, a database consisting of every scrap of data that concerns you in any way would still miss the very things that make you human and which you hold dear. Love, significance, value, and all the other "subjective" things that give our lives meaning tend to disappear into the background when the database moves into the foreground. Is it reasonable to imagine that a human being is reducible to a collection of data?

10.8 AI May Be Intelligent, But It Is Not Conscious

We are looking for a property, if any, that distinguishes human intelligence from machine intelligence. So far we have seen that it is not intelligence per se that is the distinguishing factor. Let us then add consciousness to our investigation. What is consciousness? Well we can say what self-consciousness is. It is the ability to create a mental image of oneself. We call that image "I" and learn to give it a name, "I am John Smith." We teach our children at a very young age that they "are" their name and are thrilled when they learn to respond to it. Animals, too, are conscious and perhaps possess some elements of self-consciousness.

A generic act of consciousness would consist of a perception such as "I see the tree." Such a perception is complex but singular, that is, it is one thing. Nevertheless one can analyze it into the elements "I," "tree," and "seeing." Thus a perception involves the marrying of objective and subjective poles by means of the act of perception, the "seeing." Every perception has this dual structure. It is two but, paradoxically, also one. The perception is primary and the objective and subjective elements secondary. Some people have called this kind of situation a polarity instead of a duality.

Perception, and possibly cognition, involves the creation of such polarities. You can see, for example, that you are not just conscious in the abstract but you are always conscious of something. Consciousness requires an object yet it implicitly demands a subject. This is the key to my critique of science and mathematics, which I claim are weakly but not strongly objective. Strong objectivity would dispense with the need for a subject. The natural world is then made up of objects without the need for any subject, without mind if you will. I see the natural world as being composed of perceptual or conceptual units. These require subjectivity and thus consciousness to come into being.

Let us go back to the "I," the word we use to designate the human sense of self. Notice that there are two evolutionary steps involved here. There is the conscious sense of self and then there is the word that stands for it. Animals undoubtedly have a sense of self and the rich emotional life that accompanies it. Thus they can be anxious or depressed or fearful of death. The distinguishing feature between humans and animals is language. Animals do not have language and thus the ability to objectify their subjectivity.

Where do we place an AI system in this hierarchy? This is a controversial point. There are those who feel that human consciousness arises from neural complexity and thus it is only reasonable that when a machine network arrives at the same level of complexity as our brains, then eureka!, machines will make the leap to consciousness. In this view, complexity leads to consciousness. In my view, things work in the opposite direction. They start with consciousness which leads to intelligence and complexity. In the same way that every act of perception necessarily involves a subject, consciousness also involves subjectivity. Science and mathematics, no matter what some people wish were true, are creations of human subjectivity.

I don't believe that we have to start worrying about giving voting rights to robots because none of them, no matter how sophisticated they become, will be conscious. The consciousness of machines is a myth that comforts and excites some people. We would be better off worrying about the rights of our own citizens who are disenfranchised than worrying about the rights of "human" machines.

10.9 Is Consciousness Necessary for Intelligence?

We have seen that an AI system can be intelligent, even creative, without being conscious. So perhaps consciousness is not necessary for intelligence. This is really quite an interesting idea since it opens the door to thinking of evolution, for example, as an expressions of intelligence and creativity. This is a step towards seeing creativity and intelligence as a basic feature of reality.

The startling thing about this idea for most people is that they think of creativity and intelligence as emanating from the human brain. Even if they concede that animal behavior can be intelligent, they still see intelligence as brain-based. It seems easier for such people to imagine an intelligent AI system because of the analogy with the brain, that is, the old metaphor of the brain as the computer's hardware and the mind as the software. Consider the autonomic nervous system, which is a component of the peripheral nervous system that regulates involuntary physiologic processes including heart rate, blood pressure, respiration, digestion, and sexual arousal. As the regulator of such complex systems, it is surely intelligent but lives and works for us below the level of consciousness. Life would be impossible for us if we had to consciously attend to everything of importance in our bodies. There is just too much going on. Just like a complex computer system, most of our lives take place in the dark.

Intelligence and even creativity constitute a basic dimension of the real. Maybe the mistake is assuming that creative intelligence must always be personal and brain-based. What is the alternative? It is the view that the brain is an instrument which allows us to access the creative awareness in which it is immersed. If this were the case then we would have to revise our view of the relationship between consciousness and creative intelligence.

10.10 The Non-Conscious and AI

A new perspective from which to think about AI involves including non-conscious elements in our understanding of intelligence, both of the creative and the artificial varieties. The structure of the brain and the nervous system indicates the presence of both conscious and non-conscious elements. In the first place we have the autonomic nervous system, which does its own thing below the level of consciousness. We mostly become conscious of it when it does not work properly, for example, when we develop high blood pressure. Then we need to bring it to consciousness so that we can apply corrective action.

But consider the structure of the brain itself. The number of possible pathways in the brain is absolutely enormous, effectively infinite. Yet consciousness deals with one of these neural networks at a time. By far, the majority of these neural pathways will never be used. Yet they sit there in a latent state, ready to be used if needed. This accounts for brain plasticity. When one network is down due to damage or disease, the brain can shift the damaged function to another network, or establish a new network to carry the load. The set of potential sub-networks of the brain acts as a kind of non-conscious reservoir. This explains the role of serendipity in creativity where an idea suddenly pops into your head. A new network, the physical correlate of the idea, has just been formed.

We have seen that an AI system also has an analogy to the non-conscious. AI systems depend on enormous databases which are essential and growing all the time. That is why ChatGPT wants its system to have access to all New York Times articles past and present (and others of course). Without such a vast reservoir of data the AI system does not function. Maintaining and accessing these huge databases will require enormous energy resources, which explains why the companies that are looking to manage the coming AI revolution are busy today securing these resources. Nevertheless, the database on the one hand and the output on the other give modern AI systems a deep correlation to both the brain (on which they are explicitly modeled) and the conscious/non-conscious aspects of the mind.

10.11 Creative Intelligence and AI

Creative intelligence, which one could also call knowing, is primordial. The most basic form of Mahayana Buddhism concerns the unity of being and creative intelligence, which they call form and emptiness. This primordial intelligence finds outlets in the conscious activity of human beings where it is called creativity and what I called deep thinking. However it also can be observed in non-conscious activity, either localized within a biological entity or spread out over multiple entities such as one finds in evolution, or as we have seen, emerging in artificial systems such as AI.

It, that is creative intelligence, is always operating in the background but most of the time it is hidden from view. Every once in a while, it emerges dramatically in the work of certain scientific or artistic geniuses. When we come into contact with such manifestations of creative intelligence, we are amazed and try, unsuccessfully, to pin it down, to bottle it so to speak, in order to be able to use it on command. Of course, the more we try to capture it and devise a formula for it, the further we are separated from legitimate creativity.

This explains some of our ambivalence towards AI systems. On the one hand they seem capable of intelligence and creativity, as in the case of the entanglement experiments I mentioned above. On the other, they seem to be the antithesis of the freedom that is inherent in creative expression. If an AI system is guided by an external intelligence and humanistic impulses, it holds promise of being of enormous benefit. Yet there is a dark side and this involves the possible use of AI as an instrument of totalitarian control. In this we are warned by the universal myths of the golem, the human creation that runs amok and turns against humanity.

In my writing I have often spoken of the "myth of reason" but maybe it should have been the "dream of consciousness." This is a dream of escaping from the exigencies of the human condition through the agency of some technology or way of using the mind. It is an individual dream of control in our own lives but also the collective dream of control in a culture or civilization; for example, the idea that nature only exists to benefit human beings. AI is but the latest version of this dream. AI may be fascinating but the dream is fundamentally flawed because total control is an illusion. The reason for this is that the controlling agent, ultimately the "I," is simultaneously that which we are trying to control, and that which is doing the controlling, a self-referential situation which always has the potential of producing toxic side-effects. I need only refer to atomic power and the bomb or the industrial revolution and global warming.

It may not be obvious but the flaw involves conflating the mind with thought, the contents of the mind. Let us go back to the fundamental question about any thought or idea, “Where does it come from?” It comes from mind which is, for the greater part, non-conscious. But we don’t see where it comes from; we just see the output, namely, the thought or idea. We do not see where it comes form and we *cannot* see where it comes from, for this would involve observing our own subjectivity which would, of course, make it objective. We cannot force the non-conscious to become conscious. We can only respect its existence and give it the space that will allow it to emerge.

But let me go back to the feeling of amazement we feel when some new idea is born. Why are we amazed? It is perhaps because the new idea seems so obvious and the process is so intimate and so familiar. And this feeling of intimacy arises because the new emerges from our own minds. We must distinguish, as I said above, between the process and the results of the process, between foreground and background, if you will, or between mind and the contents of the mind. Now the mystery of human existence is that “I” am both the background and the foreground, both the conscious idea and the non-conscious womb from which it emerges. The feeling of intimacy and amazement both comes from recognizing that the non-conscious self is “me.” Its claim to be me can be experienced as even deeper and more immediate than the claim of the conscious “I.”

Sometimes, using an AI system evokes these same feelings of amazement and intimacy as it reflects material back to our conscious minds. Having thus brought something to our attention, we can then work with it and integrate it with other thoughts and ideas. Then we can iterate the process as many times as it takes. It ends when we can say, “Ah, yes, that’s what I want to say.”

10.12 A Last Word

What is the most important issue that is raised by the advent of AI systems? In my opinion it is whether the advent of AI helps human beings access their own inner resources of creativity and intelligence. Does it make us more intelligent and creative or less? Like many previous technological breakthroughs, the question is whether humanity will use this new tool wisely or not. If the past is anything to go by, there is a huge danger that the dangers will be ignored. The potential economic benefits are too vast and you can already see the rush to market before any regulatory guidelines are in place.

These dangers should not blind us to the potential that certainly exists. We can use this technology to help us access our own innate ability to learn and create. Thus we can project our own creative intelligence onto such systems. But there is and there must always be a gulf between the projection, that is, the output of such systems and the creative intelligence that produced them. The creative intelligence we ultimately seek does not live in any system. It resides in us, in you, and me! Anything that encourages the development of human creative possibilities is good.

Thus, the danger is not that systems like ChatGPT are not good or not valuable, but that such systems are too good! The temptation will be to imagine that intelligence and creativity, even consciousness, reside inside the AI system. They do not. If they reside anywhere, it is in the vast potential of mind, writ large, which miraculously is accessible to every human being.

Chapter 11

Creative Intelligence and the Mind

11.1 Introduction: The Primordial Mind

This book consists of my attempt to bring the notion of creative intelligence to people's attention. This final chapter consists of a summing up, a last word. So let me begin by recalling Einstein's pregnant phrase, "the mind that is revealed in the world" that I introduced in Chapter 5. What did he mean by this? For Einstein this phrase was a succinct description of what he had learned from his lifelong meditation on the natural world. The world is not a chaotic collection of random events. It is a coherent unity. It makes sense. Because it fits together in such an intricate and unexpected fashion you can't help but feel that there is intelligence at work. This intelligence does not necessarily belong to some being or god. It comes with the territory, so to speak. Further, this intelligence is dynamic. It is creative. This creative intelligence, this coherence, is what Einstein means by "mind." So his term "mind" and my term "creative intelligence" are equivalent attempts to point to a primordial intelligence that pervades reality.

The world's mind is, of course, not the mind that we think of as residing inside an individual brain. It is a mind that resides outside of, or better, independent of, any individual's mind. And yet the extraordinary thing for Einstein (and for me) is that this intelligence is accessible to humanity—to him, to you, and to me. The mind that is revealed in the world is isomorphic to the mind that makes any conceptual system come alive, that is, the human mind. If, like Einstein, you are sensitive to such things, this is a source of the deepest mystery, the source, as well, of a sense of wonder and amazement. When one encounters this mystery it is a revelation!

Speaking about mind is another way of speaking about deep thinking. Mind was revealed to Einstein in the functioning of the natural world but that revelation arose as a result of his success in overcoming the normal point of view of his time. Einstein's intelligence lay in his ability to look at the world in a radically new way—to create a radically new conceptual system. This book is also about making a transition to a new conceptual system. Every new conceptual system is incompatible with what went before and this incompatibility is often revealed by a simple question, for example, how many numbers are there between 2 and 3? In this case the question is, "What is Mind?" The old answer is that mind comes out of a complex arrangement of neurons or possibly of computer circuits. But there is another answer that is possible and it is that mind is primordial—it does not have to come out of anything else. This statement makes absolutely no sense from our conventional point of view. If and when it comes to make sense, the world will have changed in a dramatic way. The mind of development, the mind of learning, and the mind of creating are all variants on the same basic process of creative intelligence, of deep thinking.

The core of the process that I have been describing was represented symbolically as $CS_1 \rightarrow CS_2$, the movement from a base conceptual system, developmental stage, paradigm, idea, or system of ideas to a successor system. In this chapter I shall begin by revisiting some of characteristics of deep thinking that I first listed at the end of the first chapter and then proceed to make some inferences that can be drawn from looking at things in the manner that I have described.

11.2 Mind is Natural

"Mind is natural," means that because we live in an ocean of creative intelligence the capacity for deep thinking is always present even when it may seem to be inaccessible. The potential always exists. Thus mind is not separate from the natural world. It manifests itself both in the evolution of the physical and biological world as well as in learning, development, and acts of creativity.

The primordial nature of the mind means, in particular, that there is no knowledge that is independent of a conceptual system. Mind does not come into existence at a certain moment so it is never a blank slate; development and evolution always start somewhere.

The movement $CS_1 \rightarrow CS_2$ is free. In retrospect CS_2 may seem inevitable but from the point of view of CS_1, CS_2 arrives on the scene as a revelation that was unpredictable *a priori*. That is why creativity often contains elements of serendipity, which seems to come out of "nowhere," that is, of course, nowhere within the old system. It is shocking and reframes the situation, And yet after the fact it may appear completely natural, that is, simple and even obvious.

11.3 Mind is Discrete

Mind is discrete, that is, the movement $CS_1 \rightarrow CS_2$ is discontinuous. This means that the change or insight happens suddenly. Mind is revealed in what could metaphorically be called "sparks of light." This is a part of the creative process whether in development, paradigm change, and even in evolutionary change. It is not true for everything that is usually included under the rubric of "learning," but is an essential part of a certain kind of learning—what I am tempted to call "authentic learning" and, in particular, of conceptual learning. If this discontinuity does not happen then it is doubtful that anything creative has occurred. Education, real education, is about inducing such discontinuous changes in the student. It follows that learning and development come in "stages" each of which is an equilibrium state and thus (relatively) stable. This is also true for scientific paradigms and for evolutionary changes.

11.4 Accessing the Natural Mind is Difficult

The movement $CS_1 \rightarrow CS_2$ is difficult because it depends on being able to break free of the inertia of one's habitual thought patterns. There is a tendency to cling to the security of what has worked, more or less well, in the past.

It is difficult *because* it involves a leap to a new point of view. This difficulty is an intrinsic part of all the varieties of change that have been discussed. Development and evolutionary change are also "hard" in this way. The former is evident, for example, when you see a baby struggle to stand or take her first steps. The latter follows from the idea that evolution necessarily involves a struggle to adapt to changing conditions. In such situations stasis is not an option—one must adapt or perish. It occurs to me that all of life involves such a struggle and that the notion of stasis, a time when the struggle is over and we can rest, is merely an illusion. Another way of putting this is that "life is dynamic, the universe itself is dynamic." Or else one could say that "life is difficult." The only choice is whether to accept and perhaps even enjoy this difficulty or to resist it and make things even more difficult.

11.4.1 How Can Deep Thinking be Natural and Difficult at the Same Time?

Recall that creativity (in Chapter 5) often involved holding two contradictory ides in the mind at the same time. This book as well is generated by two contradictory ideas. These ideas are: (1) deep thinking is natural and (2) deep thinking is difficult. The new perspective that I am pointing to in this book comes directly from confronting and resolving this impasse. How can something be natural and difficult at the same time? From the normal perspective being natural is incompatible with the kind of difficulty I have been describing. Yet there is another perspective where we come to see that the very same situation can be both natural and difficult. We could say that deep thinking is natural from the perspective of lantern consciousness, synthetic thought, and the right hemisphere. It is very difficult and even impossible from the perspective of flashlight consciousness, analytic thought, and the left hemisphere.

The mind of the individual is often not freely creative because it is blocked and obscured by clinging to familiar habit patterns and fear of the unknown. The natural mind is free but freedom can be frightening because it entails the loss of security. As Leonard Cohen says in the song *Closing Time*, "it looks like freedom but it feels like death." Given the choice many people would choose security over freedom. Freedom is desirable but it is not easy. The cost of freedom, it has been said, lies in eternal vigilance. The vigilance that is required involves resisting the tendency to fall back into old habits of thought, that is, conceptual systems that have outlived their usefulness.

If the bad news lies in its difficulty, the good news is that deep thinking is potentially accessible to everyone. It resides in the deeper layers of the mind. But we must work very hard and overcome a good deal of resistance if we desire to bring these deeper layers up to consciousness. This is hard work but it is an unusual kind of work for it involves undoing barriers that we have erected in our minds that prevent us from living in the present and being aware of what is real and immediate. Logic will not break down these barriers; non-logical means are necessary.

11.5 Analytic Thought Is a Simulation of Mind

Analytic thought is not a primary phenomena. It is not mind but a simulation of mind. Continuous change involves making computations within the operative conceptual system and thereby exploring one's conceptual system or paradigm. It is continuous labor that makes up the vast majority of work in science and in learning. Inevitably this work gives rise to the problematic elements which ultimately make the old situation untenable. But such work does not contain the kind of difficulty that is characteristic of deep thinking.

11.6 All Systems Have Limits

Mind cannot be contained in any one conceptual system and has no intrinsic limits. All systems have limits. Conceptual systems even if they are essentially infinite are inevitably limited and incomplete. The reality, which we identify with what we see through our operative conceptual system, is inevitably incomplete. There is a more basic reality, the "potentially conceptualized," if you will, that shows up eventually. How does it show up? In the contradictions and inconsistencies, the ambiguities and paradoxes that always arise within the conceptual reality. This is the kind of inference that we can draw from the incompleteness results of Gödel, for example, but also from a whole series of results in modern mathematics and science. It is what I called the "Blind Spot" in a previous book.[i]

11.7 Reasoning and the Limits to Logical Inference

Some people believe that we can escape from these limitations through the construction of some kind of super system—logical inference or algorithmic thinking, a theory of everything, or a super computer—that transcends all systems. This is a pipe dream. Logic does not stand outside of conceptual systems; it is itself a conceptual system.

Logic tends to cause the mind to rigidify in a certain sense. One is mesmerized by the solidity of the structure that one has erected. The result can be a tendency to rationalize inconsistencies or problems or even to deny their existence at all. I discussed this problem with logical inference in the chapter about hemispheric differentiation—the left hemisphere does not see elements that do not fit in with its operative organizational scheme. Logic is the left hemisphere's way of organizing the world. If we want to learn or to be creative we must give the right hemisphere some space and this involves breaking down the perfect and beautiful logical world that we have created for ourselves.

It is useful to distinguish between formal logic and the more general process of reasoning. Reasoning seems to be very basic and not just restricted to the left hemisphere nor to adults. Babies actively try to make sense, that is, reason, about their world. Reason might be defined as a way of using the mind to get objective information about the world. Formal logic, or the propositional calculus, is merely the abstraction of certain properties of more basic reasoning; it does not "capture" all possible ways of reasoning.

I discussed a more general way to think of reasoning in Chapter 7. In fact I have discussed at various places in the book a more generalized "reason" or what could be called "creative or developmental logic." This would involve the pre-formal elements of creative thinking in science such as coming up with the idea for a scientific experiment or mathematical proof. This kind of reasoning also concerns itself with establishing what is objectively valid. It is however not algorithmic and thus flies in the face of those attempts to reduce thinking to the analytic. Discovery and development are objective phenomena but they are not algorithmic.

11.8 When Systems Break Down

It would be wise to pay special attention to those situations in which established systems break down. Because of the tendency for the operative system to completely define the perceived reality of the given situation, the obsessive attachment to the status quo must be broken down before something new can appear. It is not that the initial system is wrong or that it will be negated or disappear when it is replaced by another system. Even after one has moved on to the successor system, the original system will still be around and will continue to operate in appropriate circumstances. The change is that the hold of the original system is broken. One no longer identifies it with the total situation.

The way we break down the hegemony of CS_1 is by accepting and living with its problematic elements. The ratio of five to three *cannot* be a number yet if you think of it in a certain way and perform certain kinds of calculations it appears to behave in ways that numbers behave. Working with this contradictory situation; refraining from rejecting it out of hand, one eventually arrives at a creative moment in which a new idea, in this case a new kind of number, sees the light.

11.9 Bootstrapping

Bootstrapping in development and paradigm change is an attempt to account for the mysterious, almost miraculous, nature of deep thinking.

How we learn is a question that is every bit as profound as the question, "How did life begin?" In fact, you could think of them as variations on the same question. How do we get something new, that is, how does a new level of organization appear out of something more elementary? Such situations always contain an element of unpredictability. Those who claim that they can predict it on the basis of some theory are inevitably working backswords. Given that you now know B can you make up some story that "explains" the movement from A to B. It is like predicting a stock market crash after the fact—you come up with some "explanation" but next time the crash happens for entirely different reasons.

The strength of bootstrapping as an explanation for development is that it incorporates both continuous and discontinuous elements—the computations within the old system as well as the sudden emergence of the new system. However explaining this movement has its limits and sometimes "bootstrapping" feels to me like an attempt to go beyond these limits and explain development without leaving room for the discontinuity—the fact that the old system is completely transcended by the new.

11.10 The "Disadvantages" of Learning and Development

There is a trade off that is implicit in development and learning. Who has not had this feeling, for example, about science and the technological developments that have arisen in its wake? In one sense it is exciting and wonderful; in another there are all kind of noxious side effects from overcrowding and pollution to the erosion of the individual's sense of personal meaning and happiness. My suggestion is that this trade-off is implicit in the primary developmental paradigm $CS_1 \rightarrow CS_2$.

Again I look to mathematics to provide some relatively simple examples. Consider again moving from the counting numbers up the hierarchy to the more complex and abstract numbers systems that I have discussed. We saw that each number system is an answer to the question, "What is (a) number?" Paradoxically, even though each level appears to be an abstraction and generalization of the kind of number introduced at the previous level, in another way each level successively narrows the focus. The first step along the way is the most dramatic for it makes number into something that is more or less well defined and manageable by analytic, focused consciousness. Before this it is much more mysterious and fleeting—you know it is something significant but you can't quite say what it is. At this stage number is more of a sense or intuition than it is a concrete object. Every movement towards what is explicit tends to hide the implicit; the denoted meaning hides the connoted meaning, the abstract hides the concrete. The potential is always larger in scope than the actual.

Take the move from the rational numbers to the real numbers. Here the gain and loss are striking. The rational numbers are more concrete, more immediate. The invention of the real numbers is an incredible accomplishment but real numbers are very abstract, complex, and ultimately remote from our immediate experience. So we gain something but we lose something as well.

This trade-off is evident in child development as we saw in the writings of Gopnik and McGilchrist. We lose touch with lantern awareness and the immediacy of the right hemisphere. It is almost a cliché to mourn the loss of the innocence, creativity, and vitality of childhood. This is at once a matter of individual development and of social development. In a way the ultimate task of education is to find ways to keep students in touch with the creativity that is everyone's birthright. The creative process mitigates the trade-off that I spoke of. A certain kind of education kills creativity and deadens students' learning experience. Creative education is a continual process of rebirth, of coming alive to the richness and excitement of learning.

11.11 The Correct Approach to Education

I have been presenting a point of view in which creative learning is central and basic. It is impossible to overstate its importance. Our very existence as a society and culture depends on the production of citizens who can thrive in an environment of continual change. This is the challenge of the present technological environment. We must never forget that the citizens of tomorrow are being produced in the schools of today. What kind of education are they receiving? Often, I'm afraid, it is an education for the problems of yesterday. Are they being equipped to face the problems of tomorrow, the problems of living in the absence of significant periods of equilibrium?

At its best education is an introduction to the practice of living a productive and creative life. It doesn't matter what subject you study; the important things that you learn transcend the particular subject. However I would maintain that mathematics is a subject particularly well suited to being taught from a creative perspective. This is ironic because mathematics is often badly taught by teachers who themselves were badly taught. Most people cannot even imagine what it would mean to study mathematics from a creative perspective and so settle for much less. Mathematics is in many ways the most difficult and problematic area of the curriculum and so it is there that the battle for the nature of education could most usefully be fought.

It may seem that the idea of approaching education from the point of view of conceptual development and creative learning is pie in the sky. Changing our attitudes in this way would indeed involve radical change but the stakes are commensurably enormous. Do we want to produce citizens who are cogs in a depersonalized machine and endlessly diverted by an endless stream of gadgets and entertainment? Or shall we experience ourselves as the expression of a creative intelligence and therefore see our lives as meaningful and coherent, filled with energy and purpose?

11.12 Evolution

I mentioned at the outset that it is very exciting to think that other natural processes follow the pattern that I have elucidated in this book—the process of creative learning and development. In the theory of evolution we also see conflict, struggle, and discontinuous change. We also see temporary equilibriums that break down and are reestablished at another level. In a very basic sense evolution is learning to adapt to and thrive in a changing external environment. It is a generalized form of learning and so it is natural that it obeys the same principles as development and learning in general. The implication is that the evolutionary process is itself a form of creative development and intelligence.

11.13 Creative Intelligence in a Creative Universe

Creative intelligence is the essence of mind. In the human domain, it is revealed in deep thinking and has many of the properties that I have outlined above. Intelligence, rationality itself, as I began this book by saying, involves creativity and conversely, creative activity is the expression of intelligence. Notice that intelligence would then be a process—something that you *do* not something that you *are* or something that you *have*. Of course this would imply that intelligence didn't come into the universe at the moment that human beings developed a certain brain size or complexity. And, indeed, many people now see intelligence and awareness as something that human beings share with animals and, perhaps, even with things like plants, which, of course, do not have brains. [Chamovitz, pages 167 and 170]. It would rather mean that the evolution of the brain was itself the result of intelligence. Intelligence, in this view, comes first.

This leads to a certain kind of philosophy, namely, that creativity and intelligence are built into the universe in the same way that the potential for learning is built into children. Creative intelligence is not something that is reserved for exceptional individuals but rather creative geniuses are exceptional in their ability to open themselves up to the creative forces in nature and within themselves. It is very exciting to imagine that we live in such a "creative universe." At a time when society seems to be overwhelmed by the enormity of the challenges it faces, creative intelligence is the place, perhaps the only place, that we can look for the solutions we so desperately require. In my view what we need exists all around us, but to access it, we need to accept the deep contradictions that exist within modern societies and cultures, and learn to live with them without flinching or looking away and especially without giving in to despair. By doing this we open ourselves up to the constructive and creative forces that surround us.

11.14 Why?

If the process that I am positing does occur then the natural question to ask is, "Why does it happen?" Is it the result of something more basic? The conceptual systems we have studied developed as a response to the dissonant. However there may be another factor in addition to the need to resolve the problematic. Creativity is its own reward, it has been said. Learning and development carry with them their own intrinsic satisfaction. When a baby takes it first steps how pleased she is with herself! So we are pushed from behind by the need to overcome dissonance but we are pulled forward by our need for something that could be described by the words unity, harmony, and coherence. We can discern in all of the situations that I have described a movement towards the establishment of complex unities. A generalized need to learn and to understand stands behind both the baby's unification of the elements of their world and the scientists' or mathematicians' creation of theoretical structures that unify vast areas of human thought and experience.

---

[i] Byers (2011).

References

Arendt, Hannah, (1966). *The Origins of Totalitarianism*, Harcourt Brace, New York.

Brooks, David, (2011). *The Social Animal: The Hidden Sources of love, Character, and Achievement*, Random House, New York.

Borwein, Jonathan, and Devlin, Keith, (2008). *The Computer as Crucible: An Introduction to Experimental Mathematics*, A.K. Peters/CRC Press.

Byers, William, (2007). *How Mathematicians Think: Using Ambiguity, Contradiction, and Paradox to Create Mathematics*, Princeton University Press.

Byers, William, (2011). *The Blind Spot: Science and the Crisis of Uncertainty*, Princeton University Press.

Byers, William, (1994). Dilemmas in the Teaching and Learning of Mathematics, For the Learning of Mathematics, Vol. 4, No. 1, 35–39.

Carey, Susan, (2009). *The Origin of Concepts*, Oxford University Press, New York.

Chamovitz, David, (2014). *What a Plant Knows: A Field Guide to the Senses*, Scientific American/Farrar, Straus and Giroux, New York.

Davis Philip J. and Hersh, Reuben, (1981). *The Mathematical Experience*, Birkhäuser, Boston.

Einstein, Albert, (1930). "What I Believe," *Forum and Century*, 84, October, pp. 193–194.

Eliade, Mircea, (1957). *The Sacred and the Profane: The Nature of Religion*, Harcourt Brace Jovanovich, New York.

Eliot, T.S., (1999 edition). *Four Quartets*, Faber and Faber, London.

Fitzgerald, F. Scott, (1945). *The Crack Up*, New York, New Directions.

Gardner, Howard, (1985). *Frames of Mind: The Theory of Multiple Intelligences*, Basic Books, New York.

Gopnik, Alison, (2009). *The Philosophical Baby: What Children's Minds Tell us about Truth, Love, and the Meaning of Life*, Picador, Farrar, Strauss and Giroux, New York.

Grosholz, Emily, (2007). *Representation and Productive Ambiguity in Mathematics and the Sciences*, Oxford University Press, New York.

Gray, Eddie, and Tall, David, (1994). "Duality, Ambiguity and Flexibility: A 'Proceptual' View of Simple Arithmetic." *Journal for Research in Mathematics Education* 25, No. 2, 116–140.

Hadamard, Jacques, (1954). *The Psychology of Invention in the Mathematical Field*, Dover, New York.

Hersh, Reuben, (1997). *What is Mathematics, Really?* Oxford University Press, New York.

Hersh, Reuben, (2014). *Experiencing Mathematics: What do We do, when We do Mathematics?* American Mathematical Society.

Hof, Robert D., (2013). *M. I. T. Technology Review: 10 Breakthrough Technologies 2013: Deep Learning*, April 2.

Kanigel, Robert, *The Man Who Knew Infinity: A Life of the Genius Ramanujan*, Washington Square Press, New York (1991).

Koestler, Arthur, (1964). *The Act of Creation*, Picador, Pan Books, London.

Kuhn, Thomas, (1962). *The Structure of Scientific Revolutions*, University of Chicago Press.

Low, Albert, (2002). *Creating Consciousness: A Study of Consciousness, Creativity, Evolution, and Violence*, White Cloud Press, Ashland, Oregon.

Low, Albert, (1997). *To Know Yourself*, Tuttle Publishing, North Clarendon, Vermont.

Marcus, Gary, (2012). Is "Deep Learning" a Revolution in Artificial Intelligence?" *The New Yorker*, November 25.

Markoff, John, (2012). Scientists See Promise in Deep-Learning Programs, *New York Times*, November 23.

Martin, Roger, (2009). *The Opposable Mind: Winning Through Integrative Thinking*, Harvard Business Press, Boston.

McGilchrist, Iain, (2009). *The Master and his Emissary: The Divided Brain and the Making of the Western World*, Yale University Press, New Haven and London.

Miller, Arthur I., (1984). *Imagery in Scientific Thought*, Birkhäuser, Boston.

Mitra, Sugata, (2007).
http://www.ted.com/talks/sugata_mitra_shows_how_kids_teach_themselves?

Pasternak, Boris, (2011). *Doctor Zhivago*, translated by Pevear, Richard and Volokhonsky, Larissa, Vintage Classics, Random House, New York.

Proust, Marcel, (2003). *In Search of Lost Time*, Tr. C.K. Scott Moncrieff and Terence Kilmartin, Revised by D.J. Enright, Modern Library, New York.

Poincaré, Henri, (1952). *Science and Hypothesis*, Dover, New York.

Rothenberg, Albert, (1996). The Janusian Process in Scientific Creativity, *Creativity Research Journal*, Vol. 9, Nos. 2 & 3, 207–231.

Sacks, Oliver, (1987). *The Man Who Mistook his Wife for a Hat and other Clinical Tales*, Harper & Row, New York.

Sierpinska, Anna, (1994). *Understanding in Mathematics*, The Falmer Press, Studies in Mathematics Education 2, London.

Spelke, Elizabeth, Ah Lee, Sang, Izard, Véronique, (2010), Beyond Core Knowledge: Natural Geometry, *Cognitive Science*, 1–22.

Tall, David, (2013). *How Humans Learn to Think Mathematically: Exploring the Three Worlds of Mathematics*, Cambridge University Press, New York.

Tall, D. and Vinner, S., (1981). Concept Image and Concept Definition in Mathematics with Particular Reference to Limits and Continuity. *Educational Studies in Mathematics*, 12(2), 151–169.

Thurston, William, (1990). Mathematics Education. *Notices of the American Mathematical Society* 37:844–850.

Trudeau, Richard J., (1987). *The Non-Euclidean Revolution*, Brikhäuser, Boston.

Wigner Eugene (1960). "The Unreasonable Effectiveness of Mathematics in the Natural Sciences." *Communications in Pure and Applied Mathematics* 13, No. 1.

Wilczek, Frank, (2008). *The Lightness of Being: Mass, Ether, and the Unification of Forces*, Basic Books, New York.

Index

www.ingramcontent.com/pod-product-compliance
Lightning Source LLC
LaVergne TN
LVHW010055220926
848484LV00005BA/1125
* 9 7 8 9 8 1 9 8 1 9 1 3 3 *